Newts and Salamanders of Europe

For
Rene and Eric

. . . where the newts beckon their follower must go.
Victor Twitty
Of Scientists and Salamanders

Newts and Salamanders of Europe

Richard Griffiths

Illustrated by
Bas Teunis,
Paul Benson and Samantha Elmhurst

T & A D POYSER NATURAL HISTORY

First published in 1996 by T & A D Poyser Ltd
24–28 Oval Road, London NW1 7DX

Typeset by Phoenix Photosetting, Chatham, Kent
Printed and bound in Great Britain at
the University Printing House, Cambridge

A catalogue record for this book
is available from the British Library

ISBN 0-85661-100-X

Contents

The colour plate section can be found between pages 52 and 53.

List of Colour Plates

Preface

My earliest encounter with a newt was as a schoolboy at a garden pond just outside London. The newt in question had unfortunately got its head stuck in the mesh of a submerged plastic basket. After carefully extracting the animal, which was none the worse for its experience, I proudly displayed my prize to the owner of the pond, who expressed considerable surprise at the find. He was totally unaware that his garden supported a thriving population of these amphibians (but perhaps even more surprised by my rather terse lecture on the hazards of plant baskets to aquatic fauna!). Indeed, newts and salamanders are small, retiring animals that most people rarely see. So why write a book about them?

The last book to be written in the English language which covers the same ground as the present text was published nearly thirty years ago. This was J. W. Steward's *The Tailed Amphibians of Europe.* Steward's book, together with Robert Thorn's *Les Salamandres d'Europe, d'Asie at Afrique du Nord,* which appeared a year or so earlier, are still widely referred to. Over the past quarter of a century, however, knowledge of newts and salamanders, and attitudes towards them, have changed dramatically. The tailed amphibians have become worthy of serious scientific study by ecologists and behavioural and evolutionary biologists. At the same time interest in their natural history has increased among amateur herpetologists and field naturalists. Rejuvenated activity on these fronts has contributed to increased concern regarding the conservation of this much-neglected group of animals. Much of the information stemming from this renewed interest in newts and salmanders is dispersed between formal scientific journals, popular articles, and unpublished reports and theses. This book is an attempt to synthesise this widely spread information into a format that will be useful to the scientist yet accessible to the layman. Indeed, this is the sort of book I would have liked to have had when I found my first newt.

Confining the geographical coverage of this book to the political boundaries of continental Europe would have meant excluding some species or subspecies which play an important role in the evolutionary story of newts and salamanders. Where appropriate these non-European forms have therefore been included. This means that the area covered by the book is broadly the western Palaearctic, but those species which have a predominantly Middle Eastern or Asiatic distribution (e.g. *Neurergus*) have been excluded for the sake of brevity.

Some newts and salamanders are much easier to study than others, and this has led to an inevitable imbalance in our knowledge of the different species. There can be few animals, for example, which have had their courtship as intensively studied as the Smooth Newt. At the other extreme, scientists still debate how many species of Cave Salamander (genus *Hydromantes*) exist in Europe, or what their most appropriate generic name should be. This imbalance in knowledge will no doubt be reflected in the text, but I hope that the book will serve to highlight the diversity of life-styles in newts and salamanders and stimulate new research to fill the gaps. To this end I have avoided the systematic approach taken by Thorn and Steward, and have instead organised the material according to an evolutionary and ecological framework.

This book is therefore really the work of hundreds of scientists and naturalists all over the world. Over the past few years some of these people have become valued friends and colleagues. Some time in the future I hope I will have the privilege to make the acquaintance of those I have not yet met.

I am particularly indebted to John Cloudsley-Thompson who first gave me the opportunity to study newts and salamanders seriously, and to the late Paul Gittins who sharpened my scientific perspective of these fascinating animals. It has also been a pleasure to share fieldwork experiences with Begoña Arano, Pim Arntzen, Brian Banks, Trevor Beebee, Lee Brady, Jonty Denton, Dag Dolmen, Andrea Fazi, Pauline Flack, Jim Foster, Cristina Giacoma, Leigh Gillett, Jon Harrison, Pierre Joly, Susanna Jones, Michael Klemens, Fred Slater, Pieter de Wijer, Julia and Ralph Wycherley, and Annie Zuiderwijk. Matteo Bonetti kindly translated a number of papers and theses. Pim Arntzen, Trevor Beebee, Tim Halliday, Sergius Kuzmin, Andrew Milner, Max Sparreboom and Pat Wisniewski provided useful comments on various chapters in draft form. Leigh Gillett allowed me to photograph his extensive collection of newts, and cast his critical eye over the complete text. The following people also responded generously and promptly to my requests for photographs: Franco Andreone, Jaques Castanet, Adrian Davies, Michael Franzen, Mario Garcia-Paris, Tim Halliday, Ciril Mlinar, Thomas Mutz, Max Sparreboom, Henk Strijbosch, Burkhard Thiesmeier, Miguel Vences and Pat Wisniewski. The artistic talents of Bas Teunis, Paul Benson and Samantha Elmhurst transformed my often vague ideas for pictures into illustrations which can say a lot more than mere narrative. In addition, many of the people mentioned above patiently replied to my varied and persistent requests for unpublished information and ideas. Their views have been incorporated where I considered it appropriate, but the book remains a personal perspective and I remain responsible for any errors or misconceptions that have slipped through the filtering process. At Poyser Andy Richford guided the gestation of this book with patience and fortitude. The Science and Engineering Research Council, Natural Environment Research Council and Joint Nature Conservation Committee have funded much of my own research on newts and salamanders over the past years, and English Nature and the Countryside Council for Wales granted licences for work on Crested Newts. Finally my wife Margaret has tolerated many long, damp nights in the field, and a home littered with salamander paraphernalia, beyond the call of duty.

Richard Griffiths (Canterbury, December 1995)

CHAPTER 1

Newt or Salamander?

Salamandra salamandra/Triturus vulgaris ♂

TO most people, the term 'biodiversity' conjures up an image of steamy forests teeming with tropical birds, insect life, and gaudily coloured frogs. Somewhere in the picture, there might be a newt or salamander, but it would probably be hiding in the leaf litter and be easily overlooked. Indeed, with under 400 species in total, newts and salamanders are not a particularly rich group of animals. Their relatives the frogs and toads, for example, have nearly ten times as many species. However, within this small group of relatively inconspicuous animals, there is an astonishing diversity of life-styles. Contrary to the popular perception of amphibians being animals of both water and land, many salamanders are completely terrestrial, and either lay eggs on land which hatch into young salamanders, or give birth to fully developed young. On the other side of the coin, other closely related species are completely aquatic, breathe using gills, and lead a fish-like existence. Newts and salamanders can be found in underground streams as well as at the tops of mountains; and in tropical forests as well as at the Siberian tundra. There are salamanders that burrow in the ground while others climb trees. Colour patterns are equally varied, and can range from dappled greens and browns that make the animal almost invisible against the forest floor, to striking yellow spots on a black background that warn potential enemies that here is a nasty-tasting mouthful. Between these various extremes lie many forms, all of which have moulded the basic amphibian life-style around various environmental pressures. This book is about the diversity of these life-styles.

The tailed amphibians all belong to the order *Caudata* (or *Urodela*), which is one of the three orders making up the class Amphibia (the other orders are the *Anura*, or frogs and toads, and the *Gymnophiona*, or caecilians). Traditionally, the amphibians are regarded as animals confined to damp habitats, which pass through an aquatic larval stage before transforming into a land-dwelling frog, toad, newt or salamander. However, many amphibians live in quite dry habitats and have modified this basic two-phased life cycle considerably. Strictly speaking, all tailed amphibians are 'salamanders', of which a 'newt' is just a special type. Newts are regarded as those salamanders which have a well-defined aquatic stage as adults. However, even within the newts there are few species which spend their time in water for the whole year. On the other hand, within the salamanders there are species which are quite aquatic. Although a hard-and-fast dividing line between the two types is not clear, in Europe the 'true newts' consist of a distinct group of amphibians which fall neatly into a single genus, *Triturus.* All other European tailed amphibians which do not belong to this genus can therefore be regarded as salamanders.

NEWTS AND SALAMANDERS IN FOLKLORE

Newts and salamanders have had long associations with ancient legends and mythology. The term 'salamander' can be traced back to a Greek term which means 'fire-lizard'. Although he never saw a live specimen, Aristotle was one of the first to suggest that salamanders had the remarkable ability to survive in fire. Later accounts date from Mediaeval times and further allude to the powers of the salamander in this direction. In the seventeenth and eighteenth centuries there are a number of reports of gruesome experiments performed on salamanders to demonstrate their ability to live in fire, and even extinguish the flames. Some accounts refer to how such unfortunate animals released fluids from the skin to put out the flames. In his book, *The History of Four-footed Beasts and Serpents*, published in 1607 (see Fig. 1.1), Edward Topsel reviews some of the older accounts, but points out that the evidence is rather contradictory:

> But the greatest matter in the salamander to be enquired after, is whether it can live and be nourished by fire, or whether it can pass through the fire without harm, or quench and put out the flame. Which opinions in the very relation and first hearing, do cross one another, for how can that either be nourished or live in the fire, which quencheth the same being put into it?

It was also believed that some salamanders possessed hair on the body which had fire-resistant properties. Indeed, in the Middle Ages asbestos was sometimes known as 'salamander's wool'. In reality, all amphibians possess soft, naked skins and are entirely hairless. The salamander was also regarded as being highly poisonous. Topsel quotes a French rhyme, which reads: 'If a salamander bite you, then betake you to the coffin and winding sheet'. He goes on to point out that salamander poison is 'not inferior to the poison of any other serpent . . . '. We now know that all salamanders have glands within the skin which release toxins as a natural defence measure. These secretions may cause skin irritations in humans, but are never life-threatening.

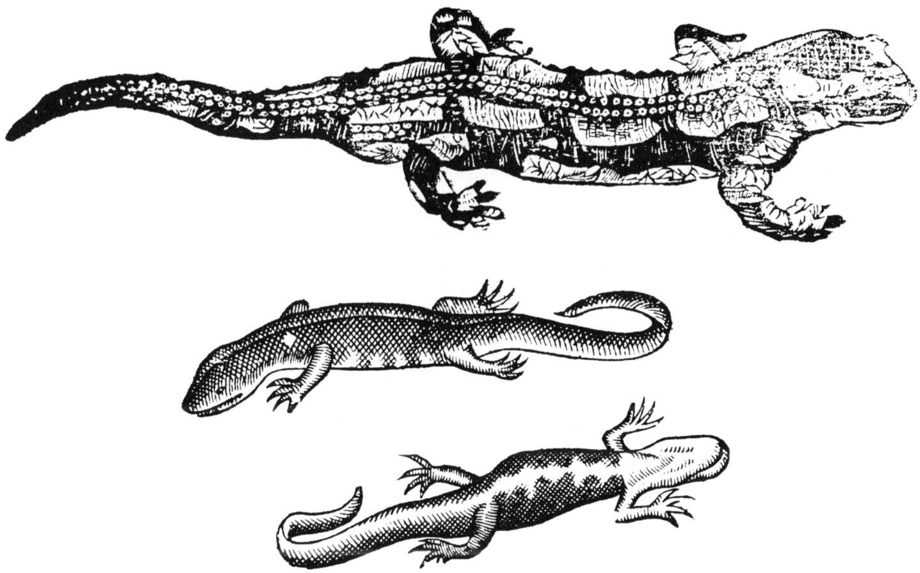

FIG 1.1 *The salamander and the newt. As illustrated in Edward Topsel's* The History of Four-footed Beasts and Serpents.

Also intriguing are reports of the salamander's feeding habits and its use in traditional medicines. Topsel states that salamanders will readily suckle cows that are lying down in woods or near hedges. Afterwards, however, the cow's milk supply will dry up, and she will be unable to produce any milk in the future. Pliny reputedly burned salamanders to ashes, and used the powder in medicines. The 'spittle' of a salamander was said to make hair fall out, and extracts from salamanders were used in potions to remove hair, hard skin and corns.

For an animal that is usually restricted to cool, moist places, and feeds exclusively on insects and other invertebrates, it is difficult to understand how some of these myths have arisen. One theory is that if logs were gathered for firewood from salamander habitats, it is quite possible that a sheltering salamander might be driven out from a crack or crevice when the log was placed on a fire. If the log was damp then the fire might go out, and the salamander would be blamed. The notion that salamanders will suckle cows could stem from the milky exudate that would be released in response to the heat or other stressful influences. If a cow lay down on a damp night and inadvertently squashed a salamander, it is quite possible that the unfortunate amphibian would release its milky fluid in its death throes. If the dead salamander was found by the farmer the following morning, he might conclude that the unfortunate beast had been killed while attempting to suckle. Whatever the origins of the mythology surrounding the salamander, this amphibian's association with fire is still reflected in the common name of Europe's most widespread species, the Fire Salamander. The term salamander is also used to describe a red-hot poker for kindling fires, and a kitchen utensil for browning meat or omelettes.

Until about 1880, newts were classified as reptiles. Before this time, the term 'newt' and its earlier derivatives were consequently used to describe either true newts or certain lizards. Indeed, Topsel describes these animals under the heading 'newt or water-lizard'. The term 'newt' seems to have had an Anglo-Saxon origin, being derived from 'evete' or 'efete'. This probably became transformed to 'ewt' in Mediaeval times, with 'an ewt' evolving to 'a newt' later on. Interestingly, the term 'eft' is still used to describe the juvenile stage of a North American species, but in recent times has assumed more general usage as a term for the recently metamorphosed stage of European newts.

Although rather less revered than the salamander, the newt has also been attributed with some strange properties throughout folklore. Pigs could apparently eat English newts with no ill effects, but in France a pig which consumed a newt was said to die. More accurate are accounts of the newt's inability to tolerate saline conditions. As Topsel wrote, 'There is nothing in nature that so much offendeth it as salt, for so soon as it is laid upon salt, it endeavoureth with all might and main to run away, for it biteth and stingeth the little beasts above measure.'

Distribution of Newts and Salamanders

Newts and salamanders are found in both the Old and New Worlds but are mainly confined to the northern hemisphere (Fig. 1.2). Although often regarded as animals of cool, temperate regions, over half of all salamander species are found in the tropical and subtropical regions of the New World. They extend below the equator in South America, but

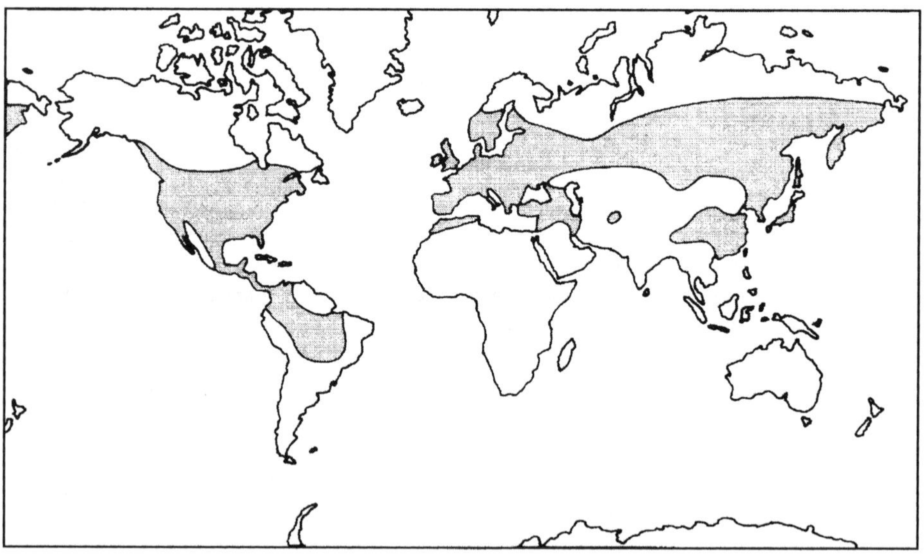

FIG 1.2 *Present day global distribution of newts and salamanders (Order: Caudata).*

only just reach northern Africa and are absent from Australia and the islands of the Indo-Malay archipelago (which otherwise support some of the highest levels of biodiversity in the world). This distribution pattern is related to barriers to dispersal which were created by continental drift or extreme climatic conditions (see Chapter 2). Had newts and salamanders reached some of those places where they are absent today, there is every reason to believe that they would have diversified as they have elsewhere.

Within the northern hemisphere, however, the range of the tailed amphibians extends from the equator to within the Arctic Circle. In many of these areas the habitat is harsh and unpredictable, but newts and salamanders have evolved intriguing life history strategies to cope with such conditions.

THE AMPHIBIAN LIFE-STYLE

Newts and salamanders, along with all other amphibians, are traditionally regarded as being incompletely adapted to living on land. Representing an intermediate stage between the aquatic fishes and the land-dwelling reptiles, their way of life is sometimes regarded as primitive and inefficient compared to the more highly evolved reptiles, birds and mammals. On first consideration, these ideas seem to be true. All newts and salamanders have a soft, naked skin which lacks the waterproof layer of the reptiles. This means that terrestrial stages are generally confined to damp habitats if desiccation is to be avoided. In addition, their method of reproduction seems to be wasteful. Many species lay several hundred eggs, of which only a tiny percentage will make it through the precarious larval stage and to metamorphosis. On top of these limitations, newts and salamanders have a body temperature which varies according to that of their surroundings (i.e. like reptiles, they are *ectothermic*). This inability to maintain a constant body temperature, as birds and mammals do, means that they become sluggish and inactive when temperatures fall. However, the idea that all of these features make the amphibians something of a retrograde group poorly adapted to any particular environment is no longer accepted. Indeed, it is precisely these features that enable them to exploit many of the habitats they occupy today.

Out of all of the vertebrates, there is one habitat type that only amphibians can exploit on a widespread basis – the temporary pool. These come in all shapes and sizes, ranging from no more than a few drops of rainwater collected within the leaves of a tropical plant to shallow dew ponds on chalk grassland (Fig. 1.3). Most fishes cannot use such water bodies because at some stage the water will disappear. In the other vertebrate groups, the development of lungs for breathing gaseous oxygen places a severe physiological restriction on living in water. Amphibians, however, are able to synchronise their two-phased life cycle to the filling and drying of small water bodies. Eggs are laid or larvae are deposited soon after ponds fill with rain, so that the young stages are able to take advantage of a rich flush of food. While the adult stages are in water for breeding activities, they are able to partially breathe and maintain body fluids through their soft, naked skins. This would not be possible if a thick waterproofing layer was present. If the water body starts to dry up, larval development may speed up so that the young emerge before desiccation occurs. Paradoxically, it may even be advantageous for some larvae to speed up growth even further by eating their siblings. This will at least ensure that some individuals will metamorphose before the pond dries, and if some have been cannibalised along the way, those 'lost' genes

FIG 1.3 *A temporary pond in mid-Wales. Although this pond regularly dries up in the summer, it supports all three species of British newts (R. Griffiths).*

may be passed to the next generation via their successful brothers and sisters! If such cannibalisation did not occur, all the larvae might stay small and none might metamorphose at all before drying occurred. Producing a large number of young, which may feed off each other later in life, is therefore all part of the amphibian way of coping with life in a temporary pool.

Having a variable body temperature also has its benefits. Maintaining a constant body temperature requires a great deal of energy, and a high rate of food intake must be sustained to enable this. Thus, compared to birds, mammals and even some reptiles, amphibians have relatively low food requirements. This means that they can live in areas where food may be sparse, and where other groups of terrestrial vertebrates would be unable to make a living.

Energy from food also fuels growth and reproduction. This means that if energy is being diverted towards maintaining a constant body temperature, there will be less available for producing offspring. With no energy being used to keep a constant body temperature, newts and salamanders are able to divert considerable resources into reproduction. This not only means that they can afford some 'wastage' in the form of egg and larval mortality, it also means that they can build up their numbers quickly when food is abundant. If there is more food, there will be more eggs and young produced, and the population will increase. If food levels fall, reproductive rate will fall and the population will decline. A variable body temperature therefore gives newts and salamanders a considerable degree of flexibility in their life-styles, and allows them to cope with fluctuations in the resources upon which they depend. When conditions are propitious, food may be abundant, and the population will grow. When conditions are harsh, food may be scarce, and the population will fall. As further discussed in Chapter 6, ups and downs may therefore be a natural characteristic of many newt and salamander populations, and simply a reflection of their ability to deal with a fluctuating and unpredictable environment.

CLASSIFICATION OF NEWTS AND SALAMANDERS

The order Caudata is divided into nine families of newts and salamanders (Table 1.1; Fig 1.4). Four of these families have representatives in Europe, although over 70% of

TABLE 1.1 *The nine families of tailed amphibians (Order: Caudata)*

Family	*No. species*	*Distribution*	*Habits*
Cryptobranchidae (Giant salamanders)	3	East and central USA, China, Japan	Totally aquatic in streams and rivers
Hynobiidae (Asiatic salamanders)	33	East and central Asia	Totally aquatic in streams and rivers
Salamandridae (Newts and European salamanders)	55	Europe, west and south-east Asia, north Africa, North America	Include pond and stream forms, but most have aquatic and terrestrial phases
Amphiumidae (Amphiumas)	3	South-east USA	Eel-like and totally aquatic with much reduced limbs
Proteidae (Olm, mudpuppy, waterdogs)	6	Slovenia, north-east Italy, east and central North America	Totally aquatic (neotenous), olm lives entirely in caves
Ambystomatidae (Mole salamanders)	35	North America	Varied group, most breed in water with terrestrial adults
Dicamptodontidae (Pacific mole salamanders)	3	Pacific coast of North America	Large salamanders, breed in water with terrestrial adults
Plethodontidae (Lungless salamanders)	210	North and South America, Italy, south-east France, Sardinia	Varied group of aquatic and terrestrial forms. Only 7 species in Europe
Sirenidae (Sirens)	3	South-east USA, north-east Mexico	Eel-like and totally aquatic (neotenous), hindlimbs absent, forelimbs reduced

species in this area belong to only one, the Salamandridae. Four of these families comprise aquatic, neotenous forms which never completely metamorphose into a terrestrial salamander (see Chapter 5). One of these families, the Cryptobranchidae, includes the largest salamanders in the world. The Japanese giant salamander (*Andrias japonicus*) can reach 1.5 m in length, and like other cryptobranchid salamanders, has folds of skin along the body which assist in the extraction of oxygen from its flowing water habitat. During breeding the eggs of cryptobranchid salamanders are fertilised outside the female's body. Members of two other families, the Amphiumidae and Sirenidae, have cylindrical, elongate bodies and live a fish-like existence in swamps, lakes or streams. The amphiumas, or as they are sometimes known 'congo eels', actually have lungs, although they retain gill slits behind the head. The limbs have become so reduced in size that they are quite useless for locomotion, although amphiumas may occasionally wriggle across land if conditions are very wet. The sirens are similar in general appearance to the amphiumas but have feathery external gills and have lost the hindlimbs altogether. Sirens are able to survive through droughts by burrowing into the mud and secreting a waterproof cocoon around the body if their pond or stream dries up. When the water body fills again the cocoon is shed and the siren swims free once again.

The fourth totally aquatic family, the Proteidae, has a single representative in Europe. This is the curious Blind Cave Salamander, or Olm (*Proteus anguinus*), which lives a com-

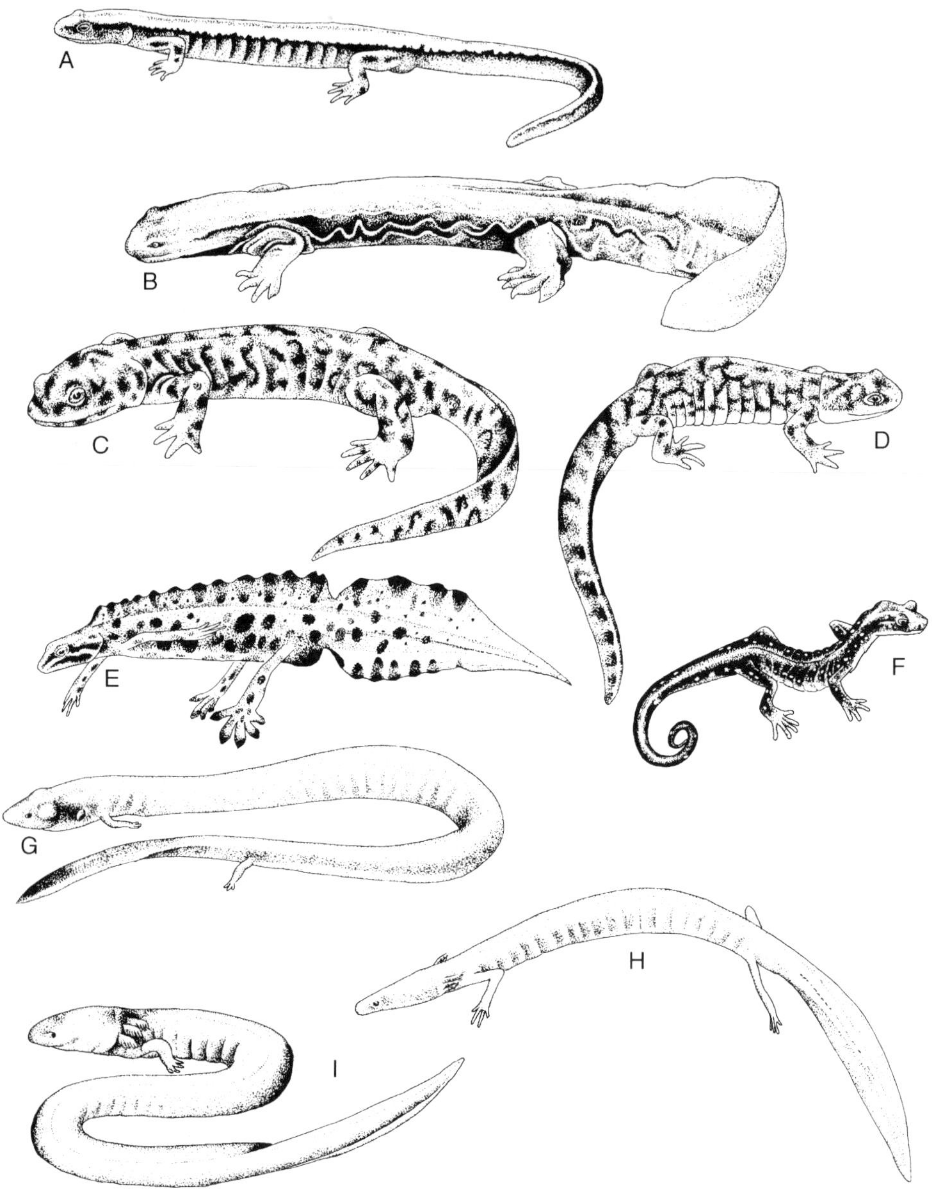

FIG 1.4 *The diversity of newts and salamanders. A, Hynobiidae* (Salamandrella keyserlingii); *B, Cryptobranchidae* (Cryptobranchus alleganiensis); *C, Dicamptodontidae* (Dicamptodon ensatus); *D, Ambystomatidae* (Ambystoma tigrinum); *E, Salamandridae* (Triturus vulgaris); *F, Plethodontidae* (Eurycea lucifuga); *G, Amphiumidae* (Amphiuma means); *H, Proteidae* (Proteus anguinus); *I, Sirenidae* (Siren lacertina). *P. Benson.*

pletely subterranean existence in the karstic areas of Slovenia and north-east Italy. Until recently, the Blind Cave Salamander was thought only to exist as a blind, pigmentless creature. However, a new form was recently discovered in south-east Slovenia which is black in colour and has well-developed eyes (Sket & Arntzen 1994). This has been designated as a new subspecies, *Proteus anguinus parkelj*. Relatives of the Blind Cave Salamander are found in eastern North America in the form of the mudpuppy and waterdogs (genus *Necturus*). These have somewhat stouter bodies than their European relative, better developed limbs, and inhabit above-ground streams and ponds. Like the Blind Cave Salamander, they have fluffy red gills richly supplied with blood vessels to absorb oxygen from the water.

The mole salamanders (family Ambystomatidae) are a diverse and abundant group in North America which display many parallels with unrelated European species. Most species of this family are classified into a single genus, *Ambystoma*. Their bodies are stoutly built, smooth skinned, and often conspicuously marked with attractive colour patterns. They live rather secretive lives in burrows, or cracks and crevices underground, and are most conspicuous after rain during the breeding seasons. Most species lay eggs in water which hatch into aquatic larvae which undergo complete development and metamorphosis. Probably the best known example of the ambystomatids is a rather exceptional one. This is the completely aquatic Axolotl (*Ambystoma mexicanum*), which has been widely used in laboratories and classrooms for research and teaching. Originally from Mexico (the term 'axolotl' has an Aztec origin), this species has been bred for many generations in captivity and exists as black, darkly spotted, or albino (i.e. pigmentless) forms. Axolotls can be induced to metamorphose into terrestrial salamanders by dosing with thyroid extract or by gradually reducing the water level in their tank. The Pacific mole salamanders (family Dicamptodontidae) are closely related to the ambystomatids, and were once classified with them in a single family. Their distribution is limited to the west coast of North America.

The Asiatic salamanders (family Hynobiidae) are a poorly studied group of small, slender-bodied amphibians, which inhabit streams and ponds. Their lungs are either reduced in size or absent. In the latter case, respiration takes place across the skin and the lining of the mouth. One species, the Siberian Salamander (*Salamandrella keyserlingii*), has a very wide range which just extends into European Russia.

By far the most diverse group of salamanders are those in the family Plethodontidae, the lungless salamanders. Like some of the Asiatic salamanders, this group relies on respiration across the skin and lining of the mouth. Despite this apparent limitation, the lungless salamanders are considered to be the most evolutionarily advanced group of tailed amphibians. This is a group of generally small, slender-bodied salamanders which are confined to damp places, but in some areas of North America can be remarkably abundant. Most species are completely terrestrial, and range from those that spend their entire lives in the surface layers of the soil, to others that can climb rocks and trees. The family also includes some aquatic forms, as well as some species which are confined to flooded caves. One of these, the Texas Blind Salamander (*Typhlomolge rathbuni*), is remarkably similar in general appearance to the Blind Cave Salamander from Europe. Despite their diversity in North and South America, only seven species are currently recognised in Europe, where they have a relict distribution in northern Italy, south-east France, and the island of Sardinia. They have largely terrestrial habits, and are confined to mountainous regions where caves and rocky outcrops provide extensive cover.

The family which has undergone the most extensive radiation in Europe is the Salamandridae. This family comprises some 12 species which are 'true newts' (genus *Triturus*), plus at least 10 European species usually referred to as salamanders. Although the salamanders are generally more terrestrial than the newts, the brook salamanders (genus *Euproctus*) and the Ribbed Salamander (also known as the Ribbed Newt, *Pleurodeles waltl*) are also highly aquatic.

The newts are conveniently divided into 'large-bodied' and 'small-bodied' forms. The large-bodied newts include the Marbled Newt (*Triturus marmoratus*), and four species collectively known as the crested newt 'species group'. These four species were formerly regarded as subspecies of just one, *Triturus cristatus*. Recent research has shown, however, that they are sufficiently distinct to warrant full species status (Bucci-Innocenti *et al.* 1983, Macgregor *et al.* 1990). The 'small-bodied' newts are at least half the size of their larger relatives, and comprise five species, including the most widely distributed salamandrid in Europe, the Smooth Newt (*Triturus vulgaris*). The Alpine Newt and the Banded Newt (*Triturus alpestris* and *T. vittatus*) are somewhat intermediate between the small and large newts, but are most conveniently grouped with the latter (Arntzen & Sparreboom 1989).

With the exception of the widespread Fire Salamander (*Salamandra salamandra*) most of the remaining members of the Salamandridae have rather restricted distributions in Europe. This factor, together with their more secretive habits, makes them rather less well studied than the newts. The European salamanders do not display the conspicuous breeding characters of male newts, and the differences between the sexes are consequently less obvious.

CHAPTER 2

Life on Land and in Water

Triturus marmoratus emerging from water

NEWTS and salamanders may live entirely on land, entirely in water, or have a life cycle which encompasses both environments. Life in these two types of environment is subject to quite different pressures, and requires different physiological and ecological adaptations. As in all animals, these adaptations are geared to ensuring that feeding, growth and reproduction occur as efficiently as possible. Newts and salamanders must therefore be able to select those habitat types where they can carry out these activities while not falling prey to other animals themselves. This chapter discusses some of the ways these goals are achieved.

THE SENSES

Vision

The eyes of newts and salamanders are generally smaller than those of frogs and toads, and primarily aquatic forms have smaller eyes than terrestrial forms (Grüsser-Cornehls & Himstedt 1976). Indeed, those species which live a permanently aquatic life in caves have much reduced eyes. In many species, though, vision is used to find food, mates and hiding places. Most terrestrial newts and salamanders will avoid strong light whenever possible, and this reaction will enable them to find dark, moist hiding places to see them through the hours of daylight when the risk of desiccation is highest. In contrast, larval stages and aquatic adults are under no risk of desiccation, and may therefore be more active in daylight.

As in other vertebrates, light entering the eye is focused on to a layer of light-sensitive cells known as the retina. The retina contains two types of light-sensitive cells, known as 'rods' and 'cones'. Rods are used in dim light, while cones are sensitive to bright light and colours. The relative proportions of these two types of cell in the eye are therefore related to day- or night-time activity. Within the newts and salamanders, the percentage of rods increases with the amount of nocturnal activity shown by different species. Smooth Newts (*Triturus vulgaris*), for example, are active during the twilight hours and only 13% of the light-sensitive cells are rods. Alpine Newts (*Triturus alpestris*) are slightly more dark-active, and have 21% rods. Crested Newts (*Triturus cristatus*) and Fire Salamanders (*Salamandra salamandra*) are exclusively active at night, and have 40% and 56% rods, respectively (Möller 1951).

Eyes which are adapted to living on land may not be suited to vision underwater. Grüsser-Cornehls & Himstedt (1976) compared the visual acuity of newts in the two environments. They concluded that when in water newts are far-sighted, while on land they are short-sighted. Feeding on land may therefore be restricted to taking prey which wander within reach, while in water prey may be detected at greater distances.

Light passing into the eye is deflected (or 'refracted') by the outer surface layer, or cornea, and by the internal lens, which focuses the image onto the light-sensitive cells of the retina. Light passing through air and water is also refracted, but to different degrees by the two media. This means that the shape of the cornea and the lens is related to the environment in which newts and salamanders live. In general, a flat cornea and rounded lens is better suited for light which is passing through water, and this type of eye is typically found in aquatic newts. Conversely, a curved cornea and flattened lens is better for focusing light which passes through air, and is therefore the type of structure present in most terrestrial salamanders (Roth 1987).

Another problem that newts and salamanders face is that in a murky pond vision is difficult, as certain wavelengths of light may be filtered out. Different pigments in the eye are sensitive to different wavelengths of light, which means that visual acuity in water may not be the same as it is on land. Rod cells in the eye contain one of two types of pigment. Pigments derived from vitamin A_2 (the porphyropsins) are sensitive to long-wavelength light, and are better suited for vision in murky water stained with organic material. In contrast, pigments based on vitamin A_1 (rhodopsins) are more suited for terrestrial vision. Newt larvae therefore have A_2 pigments in their rods, while terrestrial salamanders have A_1 pigments. Interestingly, in adult newts there is a corresponding annual rhythm of A_1–A_2 type pigments within the cells of the eye. When adults enter water for breeding, their eyes develop A_2 pigments so that they have a similar sensitivity to aquatic larvae. On leaving the pond, A_1 pigments are formed so that the eyes can function efficiently on land (Himstedt 1973).

A further question is whether newts and salamanders have colour vision. Grüsser-Cornehls & Himstedt (1976) describe some intriguing experiments in which various species were presented with different coloured artificial 'prey'. The behavioural responses towards these different colours were compared. Fire Salamanders, Alpine Newts, Crested

Newts and Smooth Newts all responded primarily to red, yellow and blue stimuli. Responses to those in the green part of the spectrum were much less frequent. It seems then, that newts, and even the dark-adapted eye of the Fire Salamander, have some capacity to distinguish colours. In newts the colours that they respond to most of all (i.e. blue and red) correspond to those used in courtship behaviour.

Further tests have elucidated the precise role of colours in the courtship of Alpine Newts (Fig. 2.1). When male Alpine Newts were presented with different coloured models of female newts, they chose the blue and red model in preference to other colour combinations (Himstedt 1979). However, the arrangement of the colours on the body

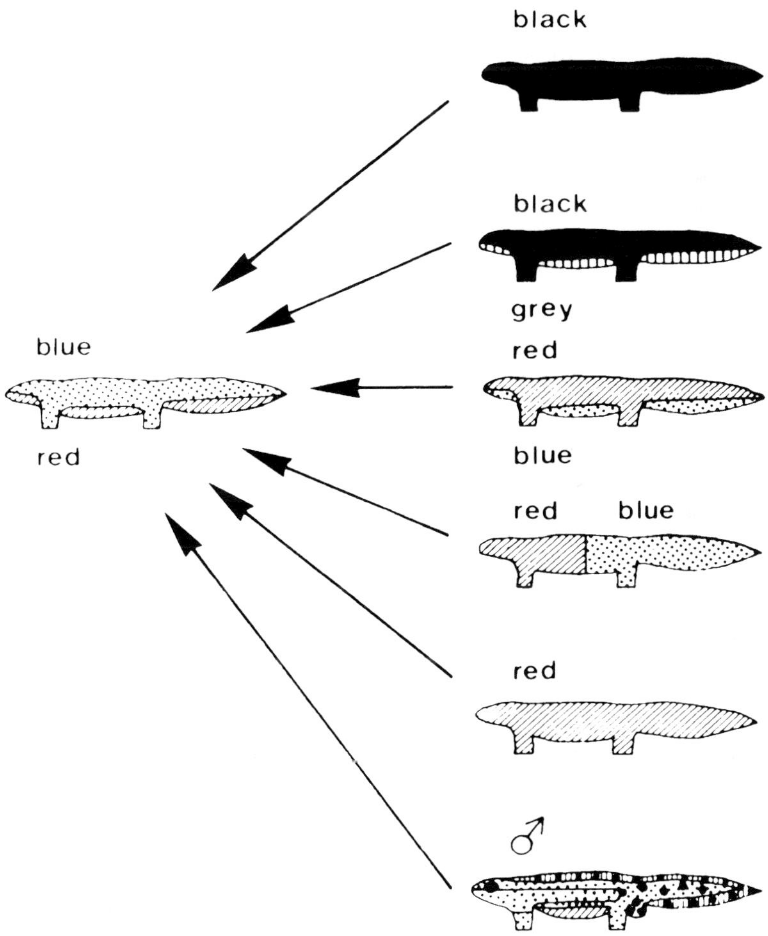

FIG. 2.1 *Colour vision in the Alpine Newt. In each test male newts were given the choice of two model 'females': a blue–red model (left) and an alternative model (right). Arrows point towards the preferred model. In all cases males preferred the 'realistic' blue–red model. From Himstedt (1979).*

were also crucial. 'Realistic' models, with blue on the upper surface and red underneath, were always selected in preference to models with other arrangements of these two colours on the body (e.g. red dorsal surface, blue belly). Colour vision may therefore be important for newts for finding both prey and suitable mates. However, feeding and courtship also rely on other cues, and these require senses other than vision for their perception.

Voice and hearing

Unlike their highly vocal frog and toad relatives, newts and salamanders do not produce sounds for the purpose of communication. Some animals may emit a short squeak if stressed or roughly handled, but this is usually an involuntary response produced by the rapid expulsion of air from the lungs. With no voice, there is consequently no need for the detection of airborne sounds. Consequently there is no external ear or eardrum. An internal ear is present, though, and this controls the sense of balance, as well as detecting vibrations. The inner ear has a muscular connection with the shoulder girdle which supports the front legs. Vibrations along the ground can be detected by transmission through the legs and shoulder girdle to the inner ear. As they do not seem to have a well-developed apparatus for detecting airborne sounds, it is widely believed that salamanders are technically 'deaf'. However, physiological measurements suggest that Fire Salamanders can detect sounds from a loudspeaker that are independent of simple vibrations. As salamanders do not use vocal communication, it is possible that such 'hearing' may help with the identification and localisation of potential enemies (Himstedt 1994).

Smell

The sense of smell is used in finding and identifying food, and in courtship behaviour. Some North American salamanders also mark territories with faecal pellets, and the sense of smell is used to identify potential competitors and their territories.

The organs of smell comprise the olfactory system, of which there are two components. These are the olfactory organ proper and the associated Jacobson's organ. (The latter organ is more highly developed in reptiles, where it is used to sense particles collected by flicks of the tongue.) The olfactory system is linked to the nostrils at the front of the head. Odours are transmitted through the nostrils by the pumping mechanism of the throat. In general, the larger the olfactory organ the more acute the sense of smell, but there is considerable variation in its importance within the newts and salamanders.

Male newts (*Triturus vulgaris* and *T. montandoni*) seem to use smell rather than vision to find mates (Cogalniceanu 1994). In laboratory choice-chamber experiments where newts were presented with a choice of male or female odours, male newts moved towards the female odours more often. However, when presented with visual cues only (in the form of a male or female newt in a sealed glass box), male newts showed no particular preference. Although these results seem different from those of Himstedt (1979) who found that Alpine Newts could distinguish mates on the basis of colour pattern (see above), Cogalniceanu (1994) points out that the sense of smell may be used to find mates at a distance, with vision being used close up. In murky ponds, however, odours would seem to be all-important at both long and short range.

Lateral line organs and electroreceptors

Newt and salamander larvae and aquatic adults possess clusters of cells in the skin which are sensitive to water currents and changes in pressure. These lateral line organs, which are also present in fishes, are distributed along the sides or back of the animals, and are particularly numerous on the head where they may be visible as distinct, dark markings. These cells are probably used to detect water movements produced by potential prey or predators. As they can only function in an aquatic medium, when newts leave the water for land the cells degenerate and no longer function.

Aquatic newts and salamanders also have organs on the head which are sensitive to electrical currents. These organs occur in shallow pits in close association with the lateral line organs, but are less numerous than the latter. These electroreceptors are thought to be used in detecting currents produced by potential prey items, such as worms, snails or water fleas. Electrical stimulation has been shown to increase the snapping behaviour of Axolotls, and larval salamanders and adult Alpine Newts may also use the electroreceptors in searching for prey (Himstedt *et al.* 1982, Himstedt 1994).

INTERNAL STRUCTURE AND FUNCTION

Skeleton

The salamander skull is a simple, somewhat flattened structure. The roof of the skull is rather open, and the orbits which hold the eyes are conspicuously large. The skull is articulated with the backbone so that the head can move up and down, and to a lesser extent, from side to side. The teeth of newts and salamanders are fairly typical of amphibians. They are small and simple in structure, with a short blunt crown connected to a longer stalk, or pedicel. Rows of teeth line the upper and lower jaws, and also occur on the palate. Teeth may break off at the joint between the crown and the pedicel but can be subsequently replaced. These simple teeth have no chewing or cutting function, they are designed merely to restrain prey in the jaws prior to swallowing. The lower jaw is connected to the upper jaw in a way that only allows vertical movement. The articulation of the skull with the backbone, and the structure of the teeth and jaws equip newts and salamanders for a 'snap, grab and swallow' method of feeding. Indeed, their diet is to a large extent dictated by the gape of the jaws, and in turn, the size of the skull: the larger the head the larger the size of prey that can be consumed.

The vertebral column of newts and salamanders serves a rather different function to that found in fishes. Living on land means that the body is no longer supported by water, and the backbone must carry the weight of the animal. This it does by transmitting the weight to the legs, via the pectoral and pelvic girdles. In salamanders which spend most of their lives on land, the limbs are short and stoutly built, as they act as load-bearing struts for the rest of the body. The ribs of newts and salamanders are much shorter than in other vertebrates. They are short, lateral processes projecting from each side of the vertebral column, but do not form a rib cage around the soft tissues of the abdomen (see Fig. 2.2). Hence there is no breast bone. The vertebral column extends from the articulation with the skull to the tip of the tail. The tail section starts from where the backbone is joined to the pelvis. The majority of newts and salamanders have a tail that is as long as, or only slightly shorter

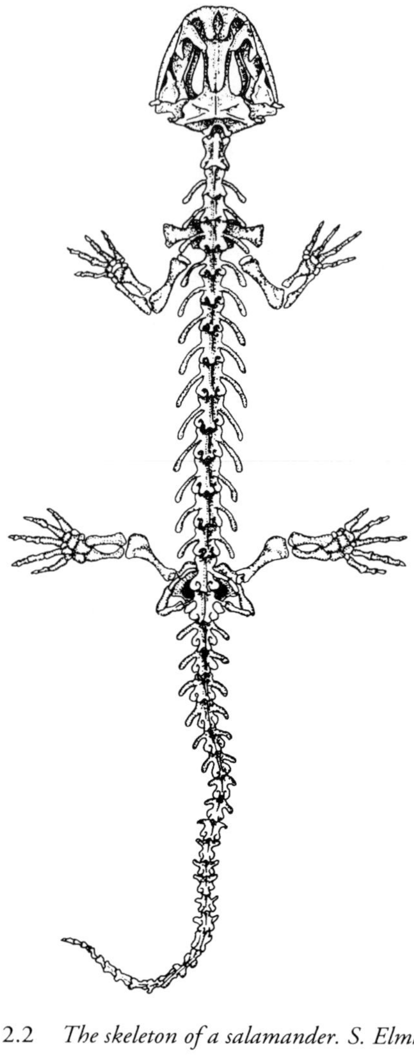

FIG 2.2 *The skeleton of a salamander. S. Elmhurst.*

than, the body. A few species, such as the Golden-striped Salamander (*Chioglossa lusitanica*) and the Caucasian Salamander (*Mertensiella caucasica*), have tails which are much longer than the body.

Front legs and back legs are similar in size, and have a bone structure similar to that of other vertebrates. There is a single thigh bone, jointed at the knee to two bones which articulate with bones of the ankle or wrist. In most species there are four digits on the front legs and five on the back. These are short structures, designed to spread the load of the weight of the body. Newts and salamanders have a remarkable ability to regenerate digits, or even whole limbs, if they are amputated. Regeneration is stimulated by electrical impulses which pass down nerves to the site of amputation, and can result in a new toe growing within a

FIG 2.3 *Female Smooth Newt with abnormally regenerated hind leg. From Griffiths (1981).*

few weeks. Sometimes injuries to legs do not heal correctly and extra limbs may grow from the point where the limb was damaged (Fig. 2.3). In one population of Smooth Newts, 4% of animals were found to have physical abnormalities arising from regenerated digits and limbs (Griffiths 1981).

Locomotion

When a newt or salamander leaves water to live on land, its muscles and skeleton must cope with quite different stresses and strains. No longer bouyed up by the surrounding medium, moving around on land is slower and uses up more energy. During locomotion on land, the body is flexed from side to side (Fig. 2.4). When the body is bent towards the right, for example, the limbs on the left side of the body are moved forward. During slow movement on a damp surface, the belly is dragged along the ground. In faster locomotion the body is entirely supported by the limbs and the tail and the belly is raised clear of the substratum. A similar method of locomotion is used when walking on the bottom of a pond. In water, however, very rapid movement is achieved by beating the tail while holding the limbs

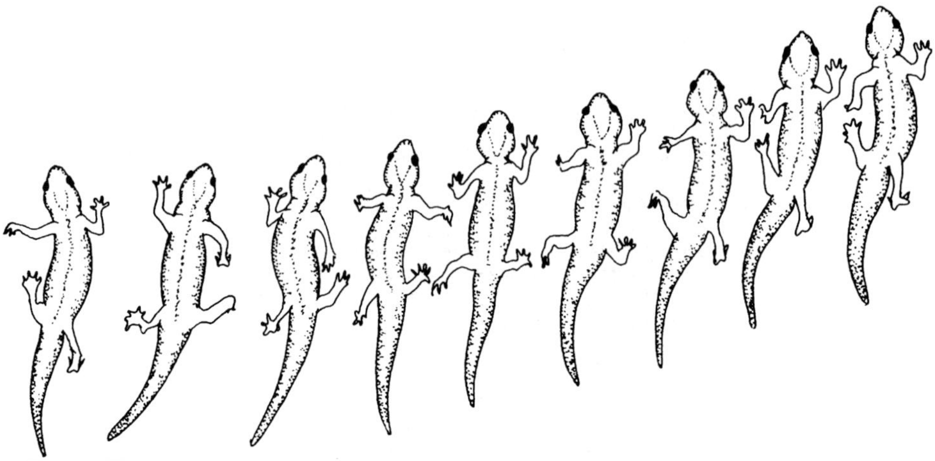

FIG 2.4 *Locomotion of a salamander on land. There is considerable lateral movement, and as the body bends towards one side the legs on the opposite side are moved forwards. Redrawn from Schaeffer (1941). S. Elmhurst.*

against the side of the body to minimise resistance. The shape of the tail therefore gives some indication of how efficient a species is in moving around in water. Tails which are flattened from side to side will provide more forward thrust when in water, and contrast with the more conically shaped tails of terrestrial salamanders. Some newts have flaps of skin or webbing on their back feet which may assist in providing a fast forward thrust, and perhaps, in braking movements. It is more likely, however, that these features are associated with courtship behaviour.

Skin

The skin of newts and salamanders performs a multitude of functions. In addition to providing an outer protective layer for the body, it also serves as a respiratory organ and regulates the inflow or outflow of water to the body. It also contains numerous glands which secrete substances to protect the body from disease and predators, and contains cells which provide a colour pattern that make the animal either highly conspicuous or blend in with the background.

In structure, the skin comprises two main layers, the outer epidermis and inner dermis, each of which may be several cells thick (Fig. 2.5). The outer layer of cells of the epidermis is called the stratum corneum, and is cornified to some degree, at least in those salamanders which spend most of their time on land. This thin layer of cells is shed from time to time as the newt or salamander grows, and then at intervals as an adult. This periodic 'sloughing' takes place either on land or in water, and begins by the old skin splitting near the head and along the back. Using muscular movements of the body coupled with rubbing against the substrate, the skin is then peeled off inside-out rather like a glove. If it is shed in one piece and floated in water, the sloughed skin may be a perfect transparent replica of its original

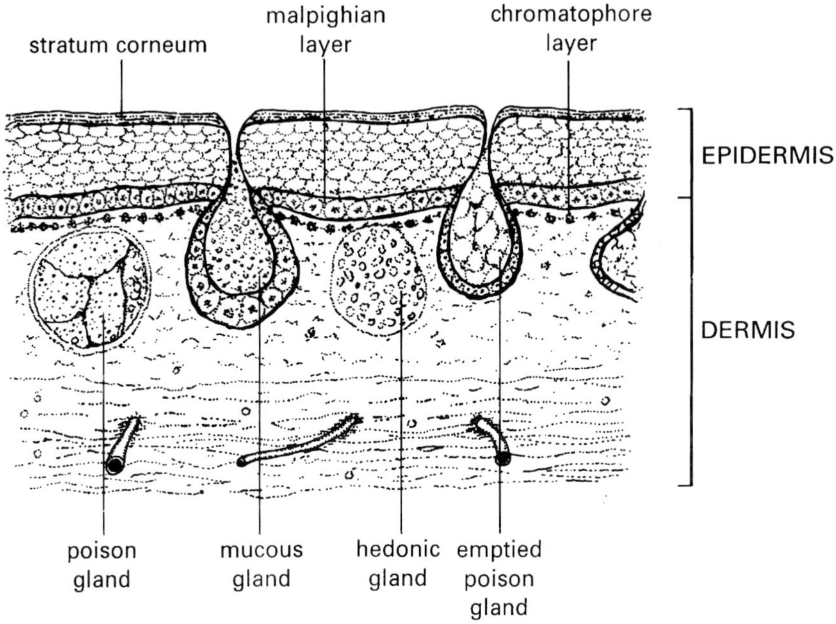

FIG 2.5 *Section through the skin of a salamander. S. Elmhurst.*

owner. More often, however, the shed skin is eaten so that no resources are wasted. Newts shed their skins before they enter water to breed in the spring, and may also do so around the time they leave later in the year. Between the stratum corneum and the dermis are several layers of cells which comprise the rest of the epidermis.

The dermis contains blood vessels, nerves and muscle fibres, as well as two types of secretory gland. Mucous glands release a colourless, slimy secretion which keeps the skin moist and protects it against infection. Granular glands secrete a milky, noxious fluid which is toxic to some potential predators. Granular glands occur only on the dorsal surface of the animal, and are usually gathered into specific areas. In the Fire Salamander and some other species, there are two conspicuous patches of granular glands behind the head, and these are termed the parotoid glands. Further rows of granular glands run along the sides, back, and on to the tail. The granular glands only release their secretions if the salamander is under stress or attacked by a predator.

Also lying within the skin are three types of cell which control pigmentation. Melanophores contain the pigment melanin, and are responsible for how light or how dark the colour of the animal is. The cells contain many branches into which melanin can disperse. When the pigment is concentrated in the middle of the cell, the animal appears light; when the pigment is dispersed throughout the cell the animal darkens. The dispersion or concentration of melanin within the cell therefore allows the animal to adjust its background colour according to environmental conditions. This effect is particularly clear in newt and salamander larvae which have numerous melanophores close to the surface of the skin. On a dark background, for example, the skin darkens, while on a light background the skin appears much paler. Xanthophores (sometimes called erythrophores) lie just beneath the epidermis, and provide yellow, orange or red colouration using a combination of pigments. The distinctive yellow spots and stripes of the Fire Salamander arise from xanthophore pigments. Very occasionally, some pigments may be missing in the cells and red-spotted salamanders result. Below the xanthophores lie the iridophores (or guanophores). These do not contain coloured pigment as such, but have a crystalline type of structure which reflects light of certain wavelengths. In combination with the xanthophores, the iridophores may bring about varied and colourful patterns. For example, the iridophores may filter certain wavelengths so that only blue light is reflected. After passing through the yellow xanthophores, this will lend a green appearance to the animal.

Respiration

All amphibians need to obtain oxygen from their surroundings, and to release the carbon dioxide which is produced as a by-product of various metabolic activities. Volume for volume, air contains around 30 times as much oxygen as water, although both media contain about the same amount of carbon dioxide. This means that aquatic salamanders have to pass much larger volumes of water over their respiratory surfaces than do terrestrial salamanders with air. A further problem for water-breathing forms is the fact that the solubility of oxygen in water depends on temperature. This means that oxygen is much scarcer in warm water than in cold water, and consequently more energy is needed to extract it.

The absorption of oxygen and release of carbon dioxide occurs across four surfaces: the lungs, the gills, the skin and the lining of the mouth. The relative importance of these four areas in respiration varies between species and according to stage of development, but many newts and salamanders will use all four at some stage during their lifetime. Although larval stages use the external gills for breathing, the lungs are developing well before metamorphosis.

The lungs consist of two sac-like structures, with many infoldings to increase the surface area over which oxygen can be absorbed and carbon dioxide released. Unlike mammals, amphibians do not have a muscular diaphragm with which to inflate and empty the lungs, and therefore use the throat to pump air in and out. The up-and-down movement of the throat pumps air through the nostrils and into the lungs. After newts enter water for breeding they must still make periodic trips to the surface to empty and refill their lungs with air. How often this occurs depends on the temperature of the water and how active the newts are. During peak activity times in the evening, the rate of 'guffing' at the surface increases (Griffiths 1985).

The gills of newts and salamanders comprise three pairs of frilly appendages behind the head. These are richly supplied with blood vessels and finely divided to increase their surface area. Gills are found in newt and salamander larvae and in fully aquatic neotenous forms, such as the Blind Cave Salamander (*Proteus anguinus*). Some species flick the gills backwards and forwards to increase the flow of water over them. In general, those species which inhabit well-oxygenated running water, such as the brook salamanders (genus *Euproctus*), have smaller gills than those frequenting more stagnant ponds.

Breathing across the skin relies on this surface being constantly moist. The lungless salamanders, which rely almost entirely on the skin for respiration, are largely confined to habitats which are continuously damp. In general, the amount of gaseous exchange that can occur across the skin increases as temperature decreases. Indeed, at very low temperatures, such as during hibernation, breathing using the lungs may not be necessary at all (Shoemaker *et al.* 1992). However, some aquatic salamanders are almost entirely skin-breathing across a wide range of temperatures (Boutilier *et al.* 1992).

To some extent the effectiveness of the skin as a respiratory organ depends on the size and shape of the salamander. Small salamanders have a greater surface area in relation to their body size than large salamanders. Similarly, long, thin salamanders have a relatively larger surface area for respiration than short, squat salamanders. The lungless salamanders consequently have small elongate bodies to maximise gaseous exchange. Even in these salamanders, however, the total surface area of the skin is much less than the total area available within the lungs for respiration (Shoemaker *et al.* 1992).

Like the gills, the lining of the mouth is richly supplied with blood vessels, and can absorb oxygen and release carbon dioxide as gaseous air passes through. Again, this surface is probably most important for respiration in the lungless salamanders, although some aquatic forms also seem able to exchange gases from water in the mouth.

How much oxygen absorption and carbon dioxide release occurs across these various respiratory surfaces also depends on environmental conditions. As body temperature varies according to the surroundings, an increase in temperature will result in an increase in metabolism, and consequently an increased demand for oxygen. To some extent, however, amphibians have an ability to shunt blood from one part of the body to another depending on environmental conditions. Under very low oxygen levels, for example, there is a risk that an aquatic salamander could lose oxygen across the gills to the surrounding water. Under such circumstances blood may be shunted away from the gills and towards the lungs (Boutilier *et al.* 1992).

Heart and blood circulation

The heart of newts and salamanders is based on a three-chamber system, comprising two atria which receive incoming blood, and a single, muscular ventricle which pumps outgoing blood around the body. Oxygenated blood returning from the lungs enters the left atrium, while the right atrium receives blood from the head, body and skin. As the two atria are connected some mixing of blood from these different parts of the body may occur here. The blood then passes into the ventricle which pumps it into arteries for circulation around the body. From a design point of view, the mixing of oxygenated and deoxygenated blood within the heart seems inefficient. Mammals and birds, for example, have a four-chambered heart which keeps the two types of blood separate and ensures that the body receives only freshly oxygenated blood. However, it is important to bear in mind that in amphibians oxygen is absorbed into the bloodstream across the skin surface as well as within the lungs. Indeed, the plethodontid salamanders have done away with lungs altogether and rely almost entirely on oxygen obtained in this way. This means that blood returning to the heart in veins from the skin will already contain a mixture of oxygenated and deoxygenated blood.

Some salamanders, such as the amphiumas and sirens, live in poorly oxygenated waters, where such mixing of oxygenated and deoxygenated blood might prove an inefficient system. These species have a more complex pumping system which can maintain a more effective separation of the two types of blood during its passage through the heart (Duellman & Trueb 1986).

The blood of amphibians contains three types of cell circulating in a fluid plasma. The most abundant cells are the red cells (erythrocytes), which are much larger than those of other vertebrates and oval-shaped with distinct nuclei. These cells contain the pigment haemoglobin, which is the oxygen carrier within the bloodstream. As water can carry much less oxygen than air, the haemoglobins in larval amphibians have a much higher affinity for oxygen than those in terrestrial forms (Duellman & Trueb 1986). This enables larvae to extract as much oxygen from the water as possible. The white cells (leucocytes) in the blood also have nuclei, and serve a defensive function. The third group of cells are the platelets (thrombocytes), which as in other vertebrates are concerned with blood clotting at the site of an injury.

Relations with the Environment

Water relations

Newts and salamanders do not drink water, but the permeable nature of the skin creates problems in terms of water balance in both aquatic and terrestrial environments. In a pond or stream, the body fluids are more concentrated than the surrounding water so there is a natural tendency for water to flow through the skin to equalise the concentrations. Conversely, on land, water will be lost by evaporation across the skin, thereby increasing the concentration of the body fluids. Newts and salamanders solve these conflicting problems with various physiological and behavioural strategies to minimise flooding of the body tissues when in water, and reduce evaporative losses when on land.

Aquatic amphibians produce large quantities of a very dilute urine, which offsets water uptake through the skin. The main excretory product is ammonia, which is highly toxic, but because it is excreted in a dilute form this presents no problems from a physiological point of view. A second mechanism is to actively pump salts into the body through the skin, to counter the dilution effect of water absorption. The third way water absorption is regulated is by hormonal control. The pituitary gland in the base of the brain releases hormones, or chemical messengers, which can directly regulate the permeability of amphibian skin and the amount of water that is released into the urine by the kidney.

Out on land, newts and salamanders actually have very little physiological control over water loss by evaporation. If a salamander is in contact with a damp substrate, water can be absorbed directly through the skin. Young Fire Salamanders, for example, can absorb water from a substrate which is only 40% saturated (Warburg & Degani 1979). Alternatively, in dry air water will be lost by evaporation. Just as the rate of oxygen absorption through the skin is related to size and shape, so is the rate of water loss in a dry atmosphere. With relatively less surface area over which evaporation can take place, large salamanders are better able to conserve water than small salamanders (Spight 1968). Similarly, a long thin salamander will expose a greater surface to evaporative water loss than a short, squat salamander. This is another reason why lungless salamanders, which rely on a large surface area for oxygen absorption, are restricted to moist habitats where there is low risk of evaporative water loss.

Although there is little difference between terrestrial and aquatic salamanders in how much water they lose through the skin, the two life-styles seem to have different tolerances of dehydration. The production of urine is much reduced in terrestrial salamanders to conserve water, and the bladder has a larger storage capacity than aquatic forms. In the Fire Salamander the bladder may hold as much as 35% of the body weight as water. Aquatic newts have much lower storage capacities, with the bladder holding only up to 5% of the body weight (Bentley 1966). In terrestrial forms ammonia is too toxic to excrete in a concentrated urine, so the main waste products are converted to urea, which is much less toxic and easily excreted in a concentrated form.

Warburg (1971) compared water loss across the skin in the Fire Salamander (*Salamandra salamandra*) and Banded Newt (*Triturus vittatus*) at different temperatures over a 3-day period (see Fig. 2.6). At low temperatures (i.e. 20°C) the newts lost less water than the salamanders. At higher temperatures (i.e. 25°C) the situation was reversed, with the newts losing more water than the salamanders. When dehydrated animals were placed in water, both species replenished most of their water rapidly, usually within 2–4 hours. However, the rehydration rate in newts was higher than in the salamanders.

Juvenile Fire Salamanders lose water more rapidly than adults, but can also rehydrate more quickly. At low humidities, however, juveniles may reduce water loss by aggregating together (Warburg & Degani 1979). This behaviour may be of benefit when ponds and their surrounding habitats start to dry up. Other behavioural tests have shown that Fire Salamanders will select substrates according to their moisture content, but that juveniles tend to select the areas of highest moisture more often (Degani & Warburg 1980).

All of these studies underline the importance of size and shape in the water relations of newts and salamanders. Stockily built salamanders, such as the Fire Salamander, may

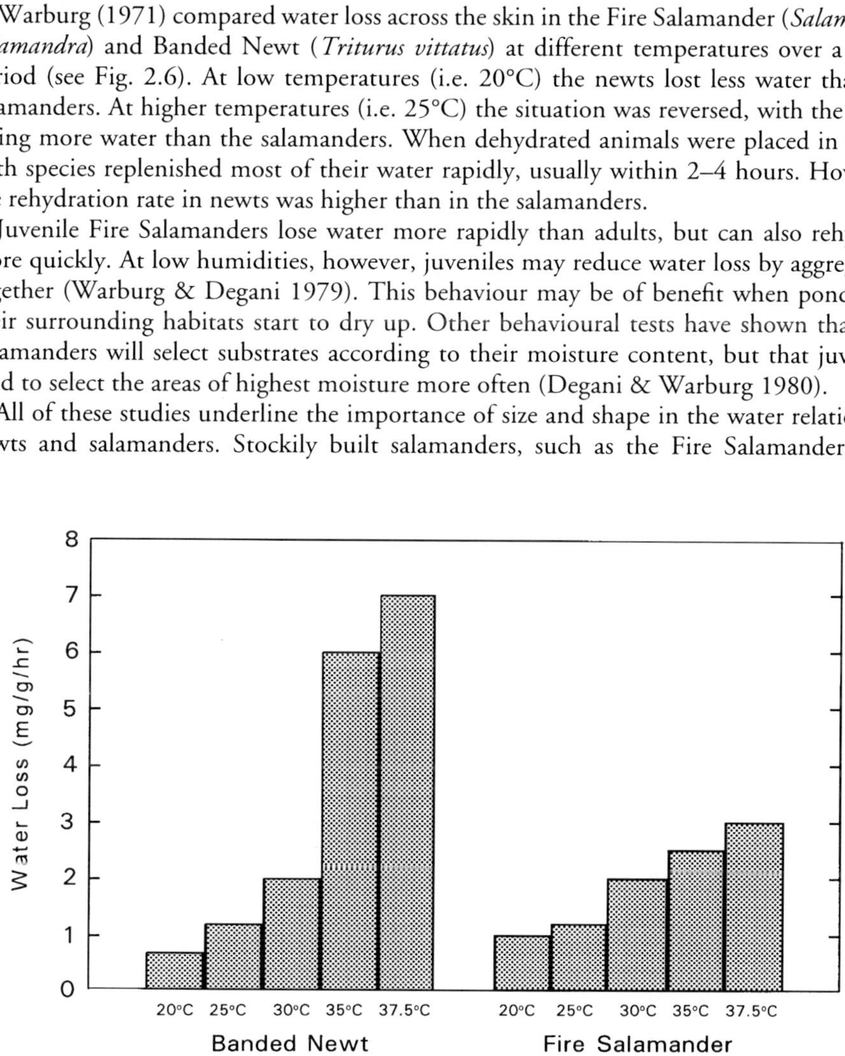

FIG 2.6 *Comparison of water loss across the skin in a newt and salamander. Animals were exposed for 4 h to dry air (0–10% relative humidity) at five different temperatures. Banded Newts lost more water at the higher temperatures than Fire Salamanders. From Warburg (1971).*

survive water loss which amounts to 40% of the body weight (Warburg & Degani 1979). The long, thin, lungless salamanders are less tolerant, however, and are likely to die if water losses amount to more than 25% of their body weight (Littleford *et al.* 1947).

Temperature regulation

Newts and salamanders, together with other amphibians and reptiles, are sometimes called 'cold-blooded'. The scientific term for this state is *poikilothermic*, which means that the body temperature varies according to environmental conditions. The blood of amphibians is therefore not always 'cold'. In fact a more meaningful term to describe how amphibians regulate their body temperature is *ectothermic*. This means that the main source of body heat is the surrounding environment (in contrast to the *endothermic* birds and mammals, which maintain a high body temperature by internal heat production).

Although the body temperatures of newts and salamanders vary according to their surroundings, this does not mean these animals have no control over temperature regulation. In its simplest form, temperature regulation may involve the avoidance of temperature extremes. Smooth newts, for example, disappear from hiding places on the surface when the temperature falls below 2°C. This is presumably due to their moving into cracks and crevices deeper underground, where the risk of freezing is reduced (Griffiths 1984). However, newts and salamanders face a problem in that the consequences of temperature regulation may conflict with other physiological requirements. Basking in the sun to raise the body temperature, for example, would result in rapid dehydration in many species. Thus the needs of water balance and respiration often override those of temperature regulation.

The body temperature of a terrestrial salamander is influenced by a number of factors (Fig. 2.7). Firstly, there may be heat gain or loss by direct convection with the surrounding

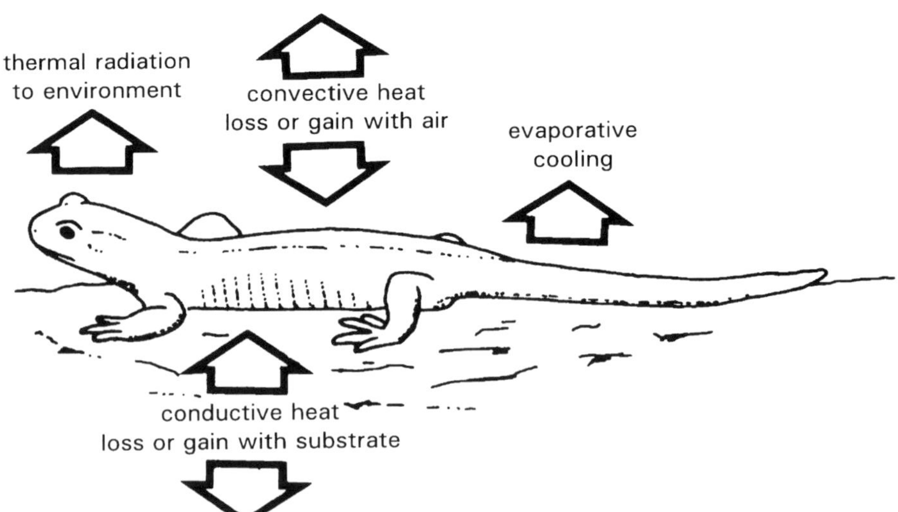

FIG 2.7 *Temperature regulation in a terrestrial salamander. Arrows show sources of heat loss and heat gain. Redrawn from Brattstrom (1979).*

air. If the surrounding atmosphere is warmer than the salamander, then the animal will lose heat, and vice versa. A similar exchange of heat will occur via conduction through the substrate. If the salamander is in a burrow or tight crack or crevice, such heat exchange will occur over much of the surface area of the body. If wandering on land, then there may be conduction between the underside of the animal and the substrate, while the rest of the body is affected by convection with the surrounding air. Two further sources of heat loss are by direct thermal radiation of body heat to the environment, and by evaporative cooling. When water is lost from the surface of the skin by evaporation, it cools the body (hence the 'sweating' response in some mammals). A salamander in a dry atmosphere will lose water by evaporation, and the body temperature may fall.

Because of the restrictions placed on temperature regulation by moisture requirements and respiration, newts and salamanders do not display many of the elaborate thermogulatory behaviours seen in many lizards and some frogs (e.g. basking in the sun, absorbing heat from warm rocks etc.). Indeed, the body temperatures of salamanders measured in the field are usually close to those of their immediate surroundings. However, tests in laboratory temperature gradients have shown that this does not necessarily mean that they are incapable of temperature regulation (Hutchison & Dupré 1992). Rather, other physiological requirements may mean that they rarely have the opportunity to do so.

In general, the temperatures at which newts and salamanders are normally active are somewhat lower than for lizards of a similar size. Feeding and courtship in newts is therefore performed at body temperatures close to that of the surrounding pond water, which in north-west Europe will vary between about 8°C and 18°C in a typical small water body in the spring. Reports of newts 'basking' at the surface of a pond on a summer's day may be more related to the presence of insect food in this part of the pond than thermoregulation. Low temperatures will not result in a complete cessation of activity, however, as there are many reports of newts actively swimming about under a layer of ice in early spring. Terrestrial salamanders are also active at relatively cool temperatures, particularly those which inhabit mountainous areas. Although there are few quantitative data, species such as the Alpine Salamander (*Salamandra atra*) and the brook salamanders (genus *Euproctus*) are difficult to keep in captivity at temperatures much above 15°C. Some salamanders can tolerate very low temperatures. Most remarkable of all is the Siberian Salamander (*Salamandrella keyserlingii*) which can survive for several weeks at −45°C. It probably manages to achieve this by replacing the water in blood and tissues with an 'antifreeze' chemical which does not form solid crystals when the temperature falls.

FEEDING

Feeding behaviour

A hungry aquatic newt will snap at a wriggling food item almost as soon as it is presented. The same animal on land will take much longer over catching its prey. In captivity, Fire Salamanders seem to be remarkably slow and inefficient at catching earthworms. Such behaviour is in contrast to the swift, darting grab the same animals use to despatch crickets. In sum, the response of newts and salamanders to potential prey is mainly visually guided, but is influenced by the size, shape and movement of the prey. Odours produced by the food also influence the feeding response, as does previous experience of different prey types. In addition, the feeding response will vary according to stage of development and on whether the animal is in an aquatic or terrestrial stage of life. Roth (1987) has summarised the sequence of behaviours that newts and salamanders use in finding and consuming prey as follows.

(1) *Orienting movement:* this is a turning of the head towards the prey, and takes place when the salamander first notices the food item.
(2) *Approaching:* the animal moves towards the prey. The distance covered by the approach movement depends on the species. Some salamanders are active hunters and can detect prey which is several centimetres away. Others are ambush predators, waiting for prey to wander by and then approaching rapidly over a short distance.
(3) *Olfactory test:* when the prey is within reach, the salamander sniffs it to check that it is palatable. Newts do this by pointing the head downwards towards the prey so that the nostrils are almost in direct contact with it. Movements of the throat may increase so that odours from the prey are pumped through the nostrils.
(4) *Fixation:* this is a final reorientation prior to snapping and varies considerably between species and according to how intensely prey food is moving.
(5) *Snapping:* the prey is secured and swallowed. In larvae and aquatic stage newts and salamanders this is usually achieved by a rapid opening of the mouth and lowering of the throat so the prey is sucked in. In terrestrial stages, the prey is seized in the jaws with the assistance of a sticky tongue. In Fire Salamanders and adult newts the tongue is hardly extruded from the mouth during prey capture. Juvenile newts, however, may flick out their short tongues to secure small prey items. The most advanced tongue-projection system is seen in the cave salamanders belonging to the genus *Hydromantes.* These salamanders can project their tongues up to a distance of 40 mm and capture prey with a high degree of accuracy even in darkness (Fig. 2.8).

These five stages of prey capture vary considerably between species, according to the type of prey being consumed, and between aquatic and terrestrial habitats. Sometimes prey capture is remarkably fast so that no distinct sequence can be detected. In other cases, prey capture may take several minutes with some stages of the sequence being repeated according to the movement patterns of the prey.

Behavioural studies on the Fire Salamander by Gerhard Roth and Gisela Luthardt-Laimer have unravelled some of the complexities surrounding prey-catching behaviour. In dim light, salamanders will snap at prey which is moving, but ignore immobile food even if it produces an appropriate odour. In total darkness, however, smell rather than vision is used, and salamanders will snap up immobile prey providing they can smell it (Luthardt & Roth 1983, Roth 1987).

Previous experience of different prey types also influences foraging behaviour, and in the Fire Salamander there seems to be a critical period in the few weeks following metamorphosis when 'imprinting' on specific prey types occurs. For example, salamanders which have only experienced non-moving prey (e.g. dead crickets) react much more readily to stationary prey later in life than those raised on moving food. Indeed, salamanders which have experienced moving prey have to rely on their sense of smell to identify an immobile food item. In contrast, those which have learnt to feed on immobile food can identify it by its size and shape alone (Luthardt-Laimer & Roth 1983).

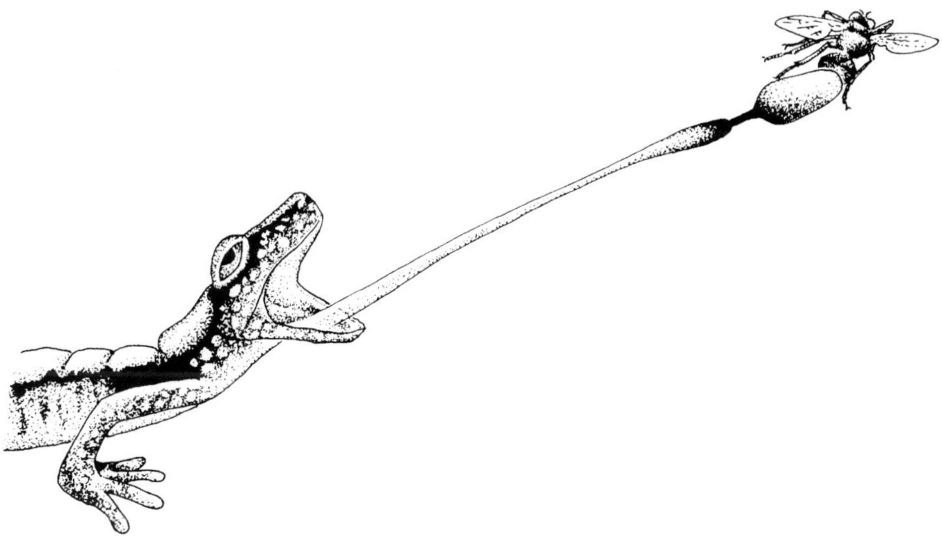

FIG 2.8 *Italian Cave Salamander catching a fly with its tongue. P. Benson.*

Even in the absence of odours, learning the appropriate sizes and shapes of suitable prey after metamorphosis is also important. When tested with odourless prey 'dummies' (i.e. black plastic shapes moved at different speeds), previous experience plays a part in food choice. Fire Salamanders raised on worms preferred long, slow-moving 'prey', while those previously fed on crickets selected, faster more compact 'prey' (Luthardt-Laimer 1983).

In an aquatic environment, the lateral line system complements smell and vision as a third mechanism for detecting prey. Experiments on Smooth Newts have shown that they are quite capable of detecting hidden food sources using non-visual cues (Ranta *et al.* 1990). When dishes of live chironomids were buried in sand in the bottom of an aquarium, newts spent more time in those areas that contained hidden chironomids than in those without food. Moreover, hungry newts spent more time near the food sources than well-fed newts. However, Smooth Newts were unable to detect hidden dishes of non-moving chironomids which had been killed beforehand. The results therefore suggest that the newts may be able to detect the movements of live chironomids beneath the sand. In natural ponds and streams, vision probably plays the all-important role in finding and catching prey, but if the water is murky newts can use their sense of smell and lateral line system.

In the total darkness of caves, other mechanisms need to be used to find prey. The larvae of the Pyrenean Brook Salamander (*Euproctus asper*) inhabit mountain streams but are sometimes found in underground passages. When offered a choice of moving or non-moving prey in light, Brook Salamander larvae preferred the moving prey, indicating that prey was detected primarily by vision and possibly the lateral line system. When the tests were repeated in darkness, however, the larvae showed no preference between moving and non-moving prey. This suggests that when underground, smell rather than vision is used as a cue for finding food (Uiblein 1992).

As they can live in either surface or subterranean streams, Brook Salamanders must maintain a flexible foraging strategy for finding food, using whichever are the most appropriate cues available within each environment. This is not the case in the Blind Cave Salamander, which lives exclusively underground. Comparisons between these two types of salamander have shown that Blind Cave Salamanders detect moving prey much more quickly than do Brook Salamanders when tested in total darkness (Uiblein *et al.* 1992).

These results show that in subterranean habitats salamanders must rely on their sense of smell and lateral line system to detect odours and movements produced by potential food items. In caves, food for salamanders is much rarer and more dispersed than it is in above-ground streams and ponds. In addition it is invisible and perhaps non-living. Living in caves therefore requires a highly efficient sensory mechanism for finding food. With a refined sense of smell and lateral line system, Blind Cave Salamanders are well adapted for foraging in this environment. Those species which do not spend their entire lives in caves have a system that allows visual detection of prey when lighting conditions permit, but the price paid for such flexibility is a less refined non-visual sensory system. So if food is scarce and foraging efficiency is low, why should species such as the Brook Salamander enter caves at all? As Uiblein *et al.* (1992) point out, cave-dwelling has advantages in that temperatures underground are more stable than at the surface, and there is a lower risk of predation in such environments.

Newt and salamander diets

In general, newts and salamanders seem to be rather unselective feeders. A wide variety of invertebrate prey is consumed, and apart from items which are too large, too small, too fast or too distasteful, food seems to be taken largely in proportion to its abundance in the field. Similarly, seasonal changes in the diet are related to annual changes in the invertebrate fauna in newt or salamander habitats. Diets are therefore varied, but are more a reflection of the variety of habitats used and the time of year than prey selection operating in different places or at different times.

Foraging theory states that animals should consume those food items which provide the best 'return' in terms of energy intake. Frequently, the most profitable prey to eat is therefore the largest that is available. However, large items of food may take longer to swallow, and in the case of a struggling worm or beetle, may actually result in the salamander expending energy in subduing its prey! Newts seem to conform to foraging theory, but only to a degree. When offered a choice of large or small water fleas, for example, Smooth Newts take more of the larger prey. However, they do not feed exclusively on such large prey, as a proportion of the smaller water fleas will always be taken whatever their abundance. In fact, large water fleas take longer to consume than small water fleas, and although they are the more profitable prey, newts are actually more successful at catching the smaller types (Ranta & Nuutinen 1985, Nuutinen & Ranta 1986). These observations are consistent with analyses of the natural diet of newts. At a pond in mid-Wales, Smooth and Crested Newts took aquatic invertebrates largely in proportion to their abundance in the pond (Fig. 2.9). However, the larger bodied Crested Newt also took larger prey than the Smooth Newt. Both species, though, consumed the smallest planktonic prey in the pond (i.e. water fleas) (Griffiths & Mylotte 1987). Larger newts

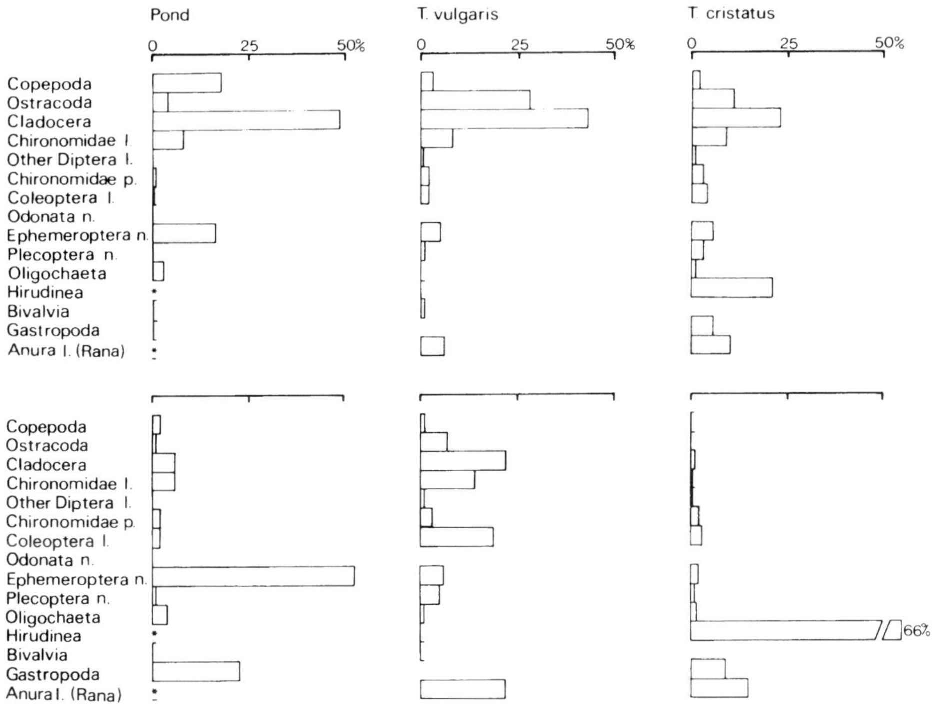

FIG 2.9 *The diet of Crested Newts and Smooth Newts displayed in relation to the relative abundance of food in a pond in mid-Wales. Figures show analyses of (a) number of prey items consumed and (b) volume of prey consumed. *indicates prey groups found in stomachs but not detected in the pond. From Griffiths & Mylotte (1987).*

may therefore have a wider diet than smaller newts, as they can feed on both large and small food items.

Although terrestrial salamanders have been traditionally viewed as consumers of soft-bodied prey, such as worms and slugs, dietary analyses have revealed a much more varied menu. Once again, on land the diet seems related to what food is actually available in the salamander's habitat. In Hungary, Szabó (1962) found that nearly 40% of the diet of Fire Salamanders consisted of soft-bodied prey (mainly slugs, with a few earthworms), while over 60% consisted of beetles, earwigs, flies, millipedes, centipedes and other hard-bodied invertebrates. The Central Spanish Fire Salamander (*Salamandra salamandra almanzoris*) lives around glacial lakes at an altitude of about 2000 m. Earthworms are extremely scarce in this habitat and the salamanders feed mainly on larval and adult beetles and flies (Guerrero *et al.* 1990). In mid-Wales, Smooth and Palmate Newts which had left water and were living on land under logs and stones, fed mainly on soil mites (Acari) and springtails (Collembolla). These two groups are probably the most abundant invertebrates in the surface layers of the soil.

Defence

Newts and salamanders have many natural enemies. Their small size and rather sluggish gait on land would seem to make them easy prey for a range of predators. However, they have evolved a number of defence mechanisms which can be quite effective at repelling an attacker.

As discussed previously, all species have granular glands in the skin which release secretions onto the skin's surface if the animal is attacked. These secretions are particularly toxic in members of the family Salamandridae, and will attack the lining of the mouth of a potential predator, and in some cases affect the nervous system. Some rather bizarre experiments performed by Miss E. Ormerod in the nineteenth century attest to the effectiveness of such secretions. After gently biting a Crested Newt, Miss Ormerod recorded her reactions with the systematic detail typical of the Victorian naturalist:

> The first effect was a bitter astringent feel in the mouth, with irritation of the upper part of the throat, numbing of the teeth more immediately holding the reptile (*sic*), and in about a minute a strong flow of clear saliva. This was accompanied by much foam and violent spasmodic action, approaching convulsions, but entirely confined to the mouth itself. The experiment was followed by headache lasting for some hours, general discomfort of the system and half an hour after by slight shivering fits. (Ormerod 1872)

In most species the release of toxic secretions is caused by the direct action of the predator's jaws squeezing the glands containing the secretion. The Fire Salamander, however, can actually spray its defensive secretions towards a predator (Brodie & Smatresk 1990). This species has the largest skin glands of any salamander, and the poison is squeezed out by the contraction of muscles around the glands running along the centre of the back. The salamander seems to adjust the direction of the spray by tilting the body towards the predator, and can achieve a spray distance of up to 200 cm. A very similar arrangement of skin glands is seen in Alpine and Luschan's Salamanders (*Salamandra atra* and *Mertensiella luschani*), so it is likely that these species can also spray chemical repellants in the same way (Brodie & Smatresk 1990). The skin toxins belong to a group of chemicals known as alkaloids, and in addition to deterring predators, they may act as broad-based antibiotics, preventing the sensitive salamander skin from attack by fungi and bacteria (Habermehl 1994).

The colour patterns of newts and salamanders vary from the cryptic, where the animal is well camouflaged, to the conspicuous, where bright colours make the animal stand out against its background. The latter colour patterns comprise contrasting shades of black, white, yellow or red. Such colours are characteristic of amphibians which are particularly toxic, and the theory is that a potential predator will quickly learn that bright colours are associated with an unpleasant mouthful. Such warning patterns are termed *aposematic* coloration.

Black and yellow patterns, typical of the Fire Salamander and many other amphibians, are therefore found throughout the animal kingdom in species which are toxic or distasteful. In some newts, the bright colouration is confined to the underside. Some species, such as the Alpine Newt, will display what is sometimes called the 'unken reflex' when attacked. This involves bending the body so that the head and tail are elevated to expose the bright red belly (see Fig. 2.10). This posture has also been observed in the Spectacled Salamander

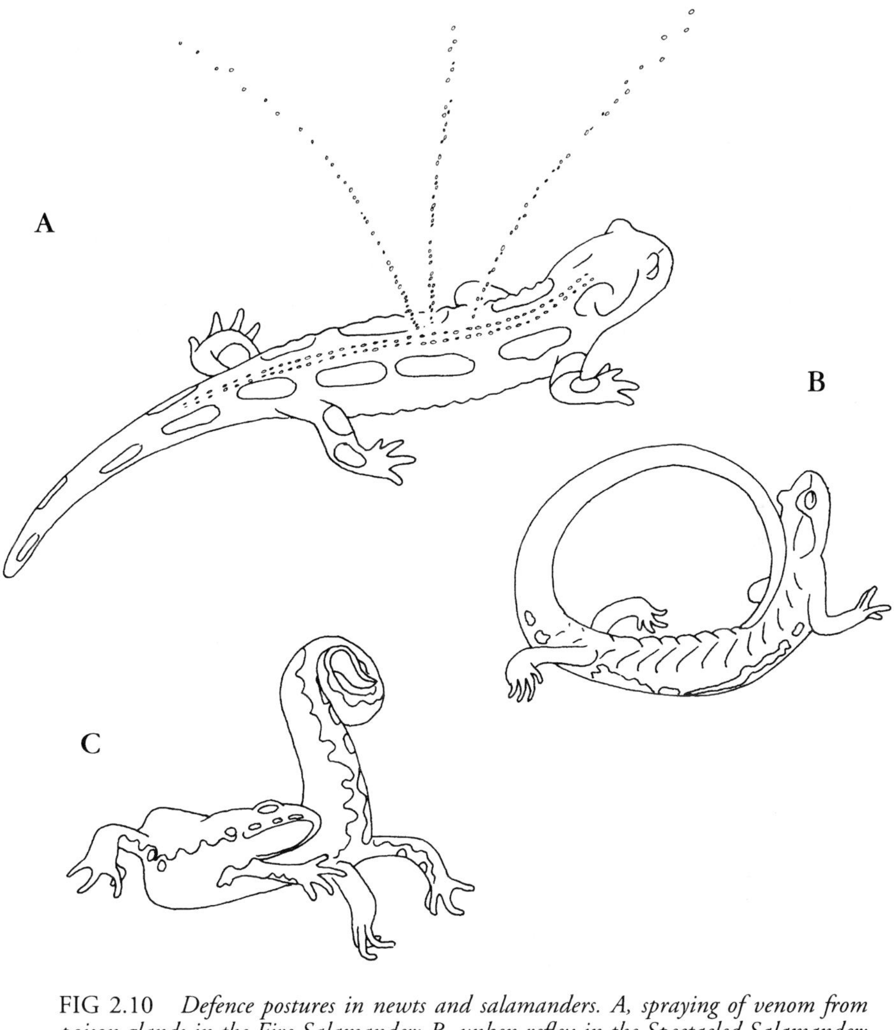

FIG 2.10 *Defence postures in newts and salamanders. A, spraying of venom from poison glands in the Fire Salamander; B, unken reflex in the Spectacled Salamander; C, unken relex with tail and body coiling in the Alpine Newt. P. Benson.*

(*Salamandrina terdigitata*) which is a drab brown on the upper surface, but exposes its strikingly marked black, red and white underside when attacked (Brodie 1983). Terrestrial-phase Crested Newts adopt a similar defensive posture when molested. This consists of closing the eyes and bending the body, with the tail curled into a tight coil. At the same time secretions from the granular glands are released and the brightly coloured belly is exposed to the predator (Andreone 1985, Denton 1990).

Many predators respond to movement of prey, and a common defence posture is just to remain immobile. Newts captured on land may maintain this posture for several minutes, and will remain immobile even if turned on their backs. This is in contrast to the violent wriggling that the same animals display when they are captured during their aquatic phase. Other defensive postures may be more aggressive, and include waving or lashing the tail. When this is done forcefully, it may serve to repel the attacker. Alternatively, it may divert the predator to the tail, which is a less vulnerable part of the body. Indeed, those species which display this posture, such as the Caucasian Salamander (*Mertensiella caucasica*), have a tail which is longer than in most other salamanders and which can be shed if the attacker is persistent.

Most salamanders have jaws that are too weak to inflict a severe bite on a predator, but the Ribbed Salamander (*Pleurodeles waltl*) from Spain is an exception as it will bite readily when attacked. This species is also unusual in that the ribs pierce the skin so that the sharp tips emerge through poison glands along each side of the body. This provides a further line of defence against a would-be attacker.

Californian lungless salamanders (*Hydromantes platycephalus*) display a most unusual escape mechanism when they are molested. Living on steep slopes, these salamanders bend the body into a tight coil, so that they roll down the hillside (Garcia-Paris & Deban 1995). Whether such behaviour also occurs in the closely related European cave salamanders remains to be seen.

CHAPTER 3

Biogeography and Evolution

Ichthyostega

STUDYING the evolutionary relationships between different types of animals is rather like trying to use detective work to solve a mystery. The various pieces of evidence must be painstakingly analysed and fitted together, but there will almost certainly remain gaps about which we can only speculate. Traditionally, species which are similar in general appearance are assumed to be closely related. By examining the bones of fossil animals, it is possible to group together those species which have certain features in common and construct an evolutionary tree. Another approach is to study living species only, and search for common features in their anatomy, physiology or behaviour. Newts, for example, have a very distinctive courtship behaviour, and those species which have similar behaviours might therefore be closely related. The problem with both approaches is that the same characters or behaviours can evolve completely independently in unrelated species, and may therefore give a misleading picture of their relationships.

In recent years, the advent of new technologies has opened up new ways of looking at evolutionary relationships. The general structure of a living organism is determined by its genes, and biochemical and molecular methods can now be used to look at the characteristics of the genes themselves. This can be done by comparing the proteins that the genes code for (using a technique called electrophoresis), or by looking at the amount and structure of the genetic material (DNA) itself. In addition the cellular bodies that carry the genetic material, the chromosomes, are themselves quite distinctive in some species, and these too can give an indication of evolutionary relationships.

Ideas have also changed in recent years regarding how data are collected, analysed and interpreted. In sum, there are now many more tools available for investigating evolutionary relationships than there were a few years ago. In many species certain methods may be easier to use than others, but the European newts are unusual in that a wide range of methods have been applied to them. The problem comes in deciding which is the best tool to use.

Before discussing the relationships between the newts and salamanders alive today, however, we shall examine some of the long-extinct amphibians, and the environmental conditions under which they evolved.

THE INVASION OF LAND

There is no doubt that the earliest four-legged vertebrates, or tetrapods, evolved from fishes sometime during the Devonian, but precisely which group of fishes were the ancestors of amphibians remains controversial. Freshwater, brackish, and marine fishes have all been proposed as amphibian ancestors at different times. Part of the problem lies in the scarcity of intermediate-stage fossils to provide a link between the ancestral fishes of the Early Devonian and the first tetrapods of the Upper Devonian. However, the debate has tended to focus on two particular groups of fishes – the lungfish (Dipnoans) and the lobe-fins (Sarcopterygians). Although opinions still differ over which of these two groups gave rise to the early amphibians, recent evidence is shifting towards the lobe-fins as the most likely candidates. Indeed, one particular group of lobe-fins, the panderichthyids, are currently arousing considerable interest as possible tetrapod ancestors. The panderichthyid lobe-fins had rather flattened bodies, and a short but straight tail. The eyes were positioned on the top of the head, suggesting that these were fishes of shallow water (Ahlberg & Milner 1994).

The earliest tetrapods appeared in the Mid-Devonian, and fossils have been recovered from locations as far apart as Scotland, Russia, Latvia and Australia. The best known genera, though, are *Acanthostega* and *Ichthyostega*, which are both from Upper Devonian deposits in Greenland. One problem with living on land rather than in water is that the body is no longer 'buoyed up' by water. This means that the weight of the body must be supported by the backbone and legs. Unlike the earlier fishes, *Ichthyostega* had sturdy legs firmly connected to a rigid backbone via supporting girdles. A well-developed rib cage was also present. Although *Ichthyostega* probably spent more time in shallow water than on land, a rib cage may have protected the internal organs from collapse under the weight of the body when out of water. Pumping air in and out of the lungs may have been achieved by movements of the rib cage, as the head seems too small to enable sufficient oxygen to have been obtained by 'throat-pumping' as in modern amphibians. A further feature which was absent in its fish ancestors was a head which could be moved from side to side. All fishes, including those existing today, have a skull which is firmly attached to the girdle supporting the front fins. This means that fishes cannot move their heads independently of their body. In *Ichthyostega* and all later amphibians, the skull is free of the pectoral girdle, which means that the head is movable on a short, but distinct, 'neck'. This freeing-up of the head was probably associated with the new 'snap and grab' method of feeding used by amphibians to take advantage of highly mobile, terrestrial insects. Some fish-like features were also retained. The short tail had vertical fin-rays, presumably supporting a fleshy tail-fin for propulsion in water. *Acanthostega* was similar to *Ichthyostega*, but more primitive. Its retention of prominent gills suggests that it was still highly aquatic.

Superficially, *Ichthyostega* resembled a large squat salamander, but it is unlikely to be a direct ancestor of today's tailed amphibians. Instead it is more likely to be an offshoot from the so far undiscovered stock which gave rise to more recent forms. Although it represented a considerable advance on its fish-like ancestors, this metre-long primitive amphibian was probably still rather clumsy when moving about on land. Indeed, *Icthyostega* was most likely a creature of the water's edge, spending most of the time in shallow water but occasionally emerging onto land to forage in moist, densely vegetated areas.

The environmental conditions which prompted the invasion of land and evolution of the first tetrapods are also controversial. What is clear is that during the Mid-Devonian, pools and swamps contained an enormous diversity of fishes, and competition and predation between them must have been intense. On land, however, there were no vertebrate animals, but an abundance of wingless insects, spiders, centipedes and millipedes, which provided a ready food source for any group ready to make the transition from water to land. Early tetrapods therefore had an opportunity to exploit a new ecological niche. This may have been the driving force behind the invasion of land and subsequent radiation of amphibians.

ANCIENT AMPHIBIANS

The first time that amphibians can be recognised as a distinct group of tetrapods is in the Lower Carboniferous (see Fig. 3.1). From this time through to the Triassic, a period covering some 13 million years, amphibians evolved into a wide range of forms, ranging in size from the 6 m long *Mastodonsaurus* to a myriad of salamander-like animals only a few centimetres long. Although the end of the reign of amphibians is often linked to competition and predation caused by the rise of the reptiles, this cannot be the whole story. Indeed, the presence of some 4000 species alive today is testimony to the success of the group in the face of pressures from 'higher' vertebrates. However, there is no doubt that amphibians were once much more diverse than they are today. Although terrestrial amphibians seemed to decline first, some large aquatic forms persisted until the Cretaceous, when they appear to have been replaced by crocodile-like reptiles or large fishes. Another theory is that amphibians declined because the climate was becoming steadily dryer. Less rainfall leads to smaller water bodies, and therefore less habitat for aquatic and larval amphibians (interestingly, similar theories have been proposed to account for the present-day declines of amphibian populations – see Chapter 7). It is also possible that some environmental catastrophe at the end of the Permian led to the extinction of many groups. Present-day amphibians may have evolved from just one or two groups that survived such a disaster. The patchy fossil record has also led to a large number of theories regarding how present-day amphibians evolved from these early forms. So is it possible at all to look back at fossil amphibians and identify the possible ancestors of today's newts and salamanders?

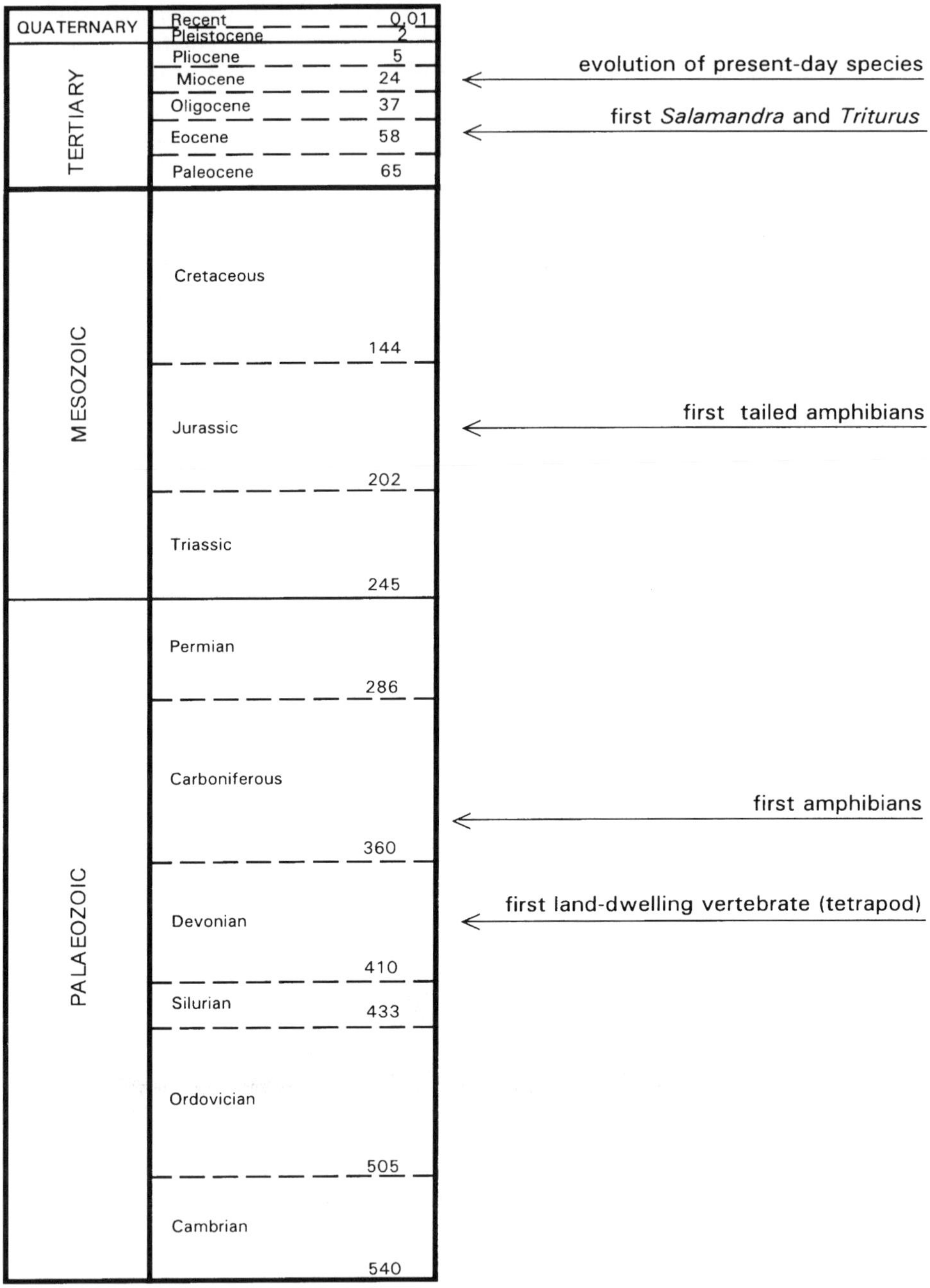

FIG 3.1 *The geological timescale, showing main events in the evolution of amphibians. Numbers indicate the start of the different periods as millions of years before present.*

One thing that most researchers agree on is that despite their superficial differences, present-day frogs, toads, newts, salamanders and caecilians are actually quite closely related. Indeed, in anatomical terms the three amphibian classes – Anura, Caudata and Gymnophiona – have more in common with each other than they do with any of the fossil forms. This means that the three amphibian groups probably share a common ancestor, which lived sometime during the Palaeozoic. Although opinions differ, there seem to be three fossil groups which could contain such a common ancestor – the nectrideans, the microsaurs and the temnospondyls (Milner 1993).

The nectrideans were elongate amphibians with long tails from the Upper Palaeozoic, and measured up to 75 cm long. (Interestingly, they had a vertebral column composed of bones which were very like those of present-day newts.) Most species probably foraged for food in water or amongst leaf litter or loose soil. Some had flattened tails to assist with swimming, but the most remarkable are those with enormous horns on either side of the head. These horns were outgrowths of the skull, and it is still unclear what purpose they served. Nectrideans have been proposed as relatives or ancestors of present-day amphibians by various authors, although Milner (1988) has recently rejected these theories.

The microsaurs were small, mainly terrestrial amphibians, some of which resembled modern salamanders in appearance. However, some researchers believe that the microsaurs are more closely related to the worm-like caecilians. The temnospondyls were a diverse group of amphibians which lasted from the late Palaeozoic through to the Triassic. A typical temnospondyl, such as *Eryops*, had short legs, and a large flat head. This suggests it was incapable of efficient locomotion on land and therefore was confined to aquatic or semi-aquatic habitats (Fig. 3.2). Other forms, such as *Cacops*, had sturdier limbs, indicative of a more terrestrial existence. With no crocodiles to compete with and an abundance of fish, some forms, such as *Trematosaurus*, became specialised fish-eaters and evolved an elongated snout. What is more remarkable, however, is that *Trematosaurus* was a marine amphibian, perhaps returning to freshwater on a seasonal basis. No living salamanders or newts can tolerate indefinite immersion in seawater.

Although some workers believe that the worm-like caecilians arose from the microsaurs sometime in the Palaeozoic, there seems to be increasing evidence that the ancestor of all

FIG 3.2 *A temnospondyl amphibian from the Permian,* Eryops. *S.Elmhurst.*

three groups of present-day amphibians was a temnospondyl of some sort (Milner 1988, 1993). One group of temnospondyls, the amphibamids, were superficially similar to modern salamanders, but had long legs and frog-like heads. They also had teeth and bones in the skull which were similar to present-day species, and included forms that entirely lacked scales (although all modern frogs, toads salamanders and newts have soft, naked skins, many fossil amphibians possessed scales). Although the evidence is shifting towards a temnospondyl ancestor for present-day newts and salamanders, the problem will not be fully resolved until more definitive fossils are found to fill the gaps in the record.

THE RISE OF THE SALAMANDERS

From the Permian until the Jurassic, the continents we know today were coalesced into a single super-continent known as Pangaea. The northern part of Pangaea consisted of present-day North America, Greenland, Europe and Asia – an area sometimes referred to as Laurasia. The southern part of Pangaea comprised Africa, South America, India, Madagascar, Antarctica, Australia and New Guinea. This vast region was known as Gondwana (Fig. 3.3). Although Laurasia and Gondwana were connected, a large sea known as the Tethys lay between the northern and southern parts of the super-continent. The earliest salamanders evolved in the northern temperate zone of Laurasia, where the climate was cool and humid. They later probably became widespread across Laurasia, but dispersal into Gondwana was prevented by the dry, tropical conditions prevailing there. The first true fossil salamanders are *Kokartus* from mid-Asia and *Marmorerpeton* from England, both from deposits laid down in the Jurassic (Milner 1993). During the Jurassic, however, Pangaea started to split up. This was accompanied by climatic upheavals and incursions of shallow seas which divided up the ranges of the early salamanders into a number of isolated groups. It was these groups that subsequently gave rise to the families we know today.

At this time North America was still connected to Europe, but a shallow sea, the Turgai, separated the joint land mass of Euro-America from much of present-day Asia. To the east of the Turgai sea evolved the hynobiid and cryptobranchid salamanders. The cryptobranchids, which include the largest species alive today, are thought to be neotenous descendants of the hynobiids. The hynobiids are still almost entirely confined to Asia, but the cryptobranchids are also found in North America. This suggests that at some stage the ancestral stock crossed from Asia into North America via a bridge of land across the Bering Straits (although an alternative theory is that the cryptobranchids evolved in North America and made the crossing in the other direction). To the west of the Turgai sea occurred several groups of salamanders which were to later give rise to the newts and salamanders of Europe and North America. The incursion of the Atlantic ocean led to a final split between North America and Europe/Asia, and divided these salamanders into two geographical groups. The European segment contained members of two families – Salamandridae and Proteidae. The family Proteidae survives with one sole member in southern Europe, the neotenous Blind Cave Salamander (*Proteus anguinus*). The Salamandridae radiated into a large number of species in western Europe, encompassing predominantly terrestrial (i.e. 'typical' salamanders) and partly aquatic (i.e. 'typical' newts) forms.

By the Upper Eocene, then, there are fossils which are clearly recognisable as belonging to the genera *Salamandra* and *Triturus* (Milner *et al.* 1982). The subsequent withdrawal of the Turgai sea which had divided Europe and Asia enabled these early salamandrids to disperse eastwards, and the Asiatic cryptobranchids to disperse westwards. Indeed, some newts made it as far as the Japanese archipelago, before this land mass split from Asia. One of the present-day Asiatic genera, *Tylototriton*, was also widespread in Europe during the Eocene and Oligocene. It is possible that the diversification of newts of the genus *Triturus* may have have led to the decline of *Tylototriton* in this continent.

Although the splitting of North America from Europe and the development and subsequent withdrawal of the Turgai sea go a long way to explaining the present distribution of salamanders and newts in Europe, some problems still remain. Confined to south-east France, northern Italy and Sardinia are seven species of lungless salamanders (family Plethodontidae), a group which is widespread in North America. How did this small group arrive in Europe and then contract in range to occupy the relict distribution pattern that it does today? The plethodontid salamanders are a group that became isolated in the New World when the Atlantic Ocean was formed, and subsequently radiated into a large number of species there. Their arrival in Europe must therefore have occurred in more recent times. This is borne out by the presence in Europe of some salamandrids which presumably evolved to fill plethodontid-type niches before the arrival of lungless salamanders. The Golden-striped Salamander (*Chioglossa lusitanica*), and Luschan's Salamander (*Mertensiella luschani*) are good examples of such convergent evolution (Wake & Özeti 1969). A widely held theory is that a forest corridor linked North America and Asia across the present-day Bering Straits sometime during the Tertiary era. The ancestors of the European plethodontid salamanders entered Asia via this route, migrated into Europe, but subsequently became extinct across a large part of their range as a result of the Pleistocene glaciations (see below). This seems the most likely route for the ancestors of the lungless salamanders to have made their way into Europe.

An alternative theory for the arrival of the plethodontid salamanders is that they arrived via a west to east route, crossing Greenland at a time when the climate of this region was more favourable. As this theory assumes that they never progressed beyond western Europe, it is not necessary to infer later extinctions to explain their absence from Asia. A third, and more speculative idea, is that the plethodontids arrived in the Eocene from the Americas, along with a range of other species native to this region. The evidence for this is based on a number of South American birds, mammals, reptiles and amphibians which turn up in the fossil record of Europe at this time. Although fossils of lungless salamanders are missing from this record, it is possible that these curious salamanders arrived by way of the same route as the higher vertebrates. The major problem lies in the nature of the land connection between South America and Europe, which, according to continental drift, did not exist at this time in geological history. How these South American fossils arrived in Europe is therefore a mystery.

The Bering land bridge which may have allowed the lungless salamanders to colonise Europe may also explain another puzzle. The typical salamanders of North America are represented by the family Ambystomatidae. Indeed, there are many parallels between this family and the Salamandridae of Europe, with different species in the two continents evolving to exploit very similar niches. The Spotted Salamander (*Ambystoma maculatum*) of North America, for example, is strikingly similar in appearance to the Fire Salamander of Europe (*Salamandra salamandra*). Again, the evolution of members of these two distinct families

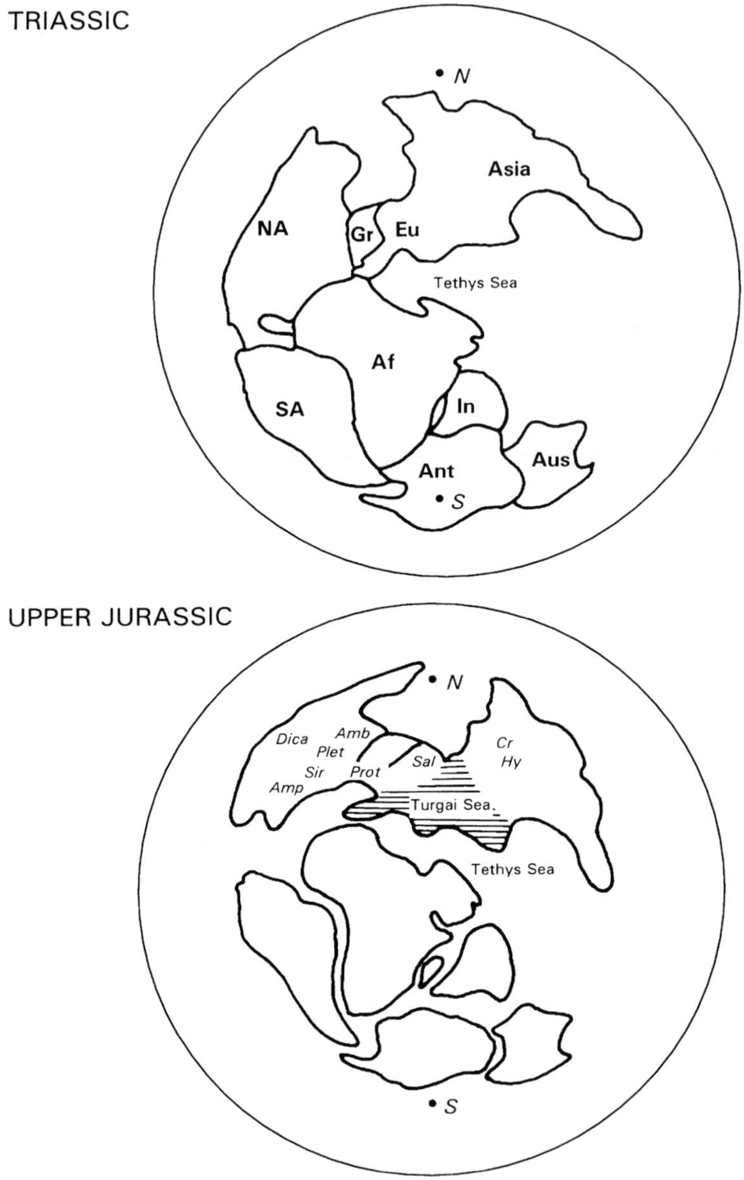

FIG 3.3 *Continental drift and its effects on the distribution of salamanders. N and S show positions of the North and South Poles, respectively. NA, North America; SA, South America; Eu, Europe; Af, Africa; Gr, Greenland; In, India; Ant, Antarctica; Aus, Australia. Possible areas for the evolution of the present-day families of newts and salamanders are shown: Amb, Ambystomatidae; Dica, Dicamptodontidae; Plet, Plethodontidae; Sir, Sirenidae; Amp, Amphiumidae; Sal, Salamandridae; Prot, Proteidae; Cr, Cryptobranchidae; Hy, Hynobiidae. Triassic: Laurasia and Gondwana*

UPPER CRETACEOUS

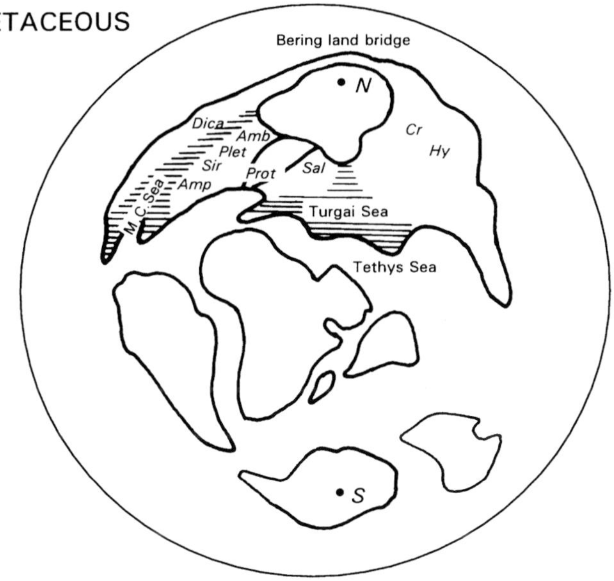

PALAEOCENE

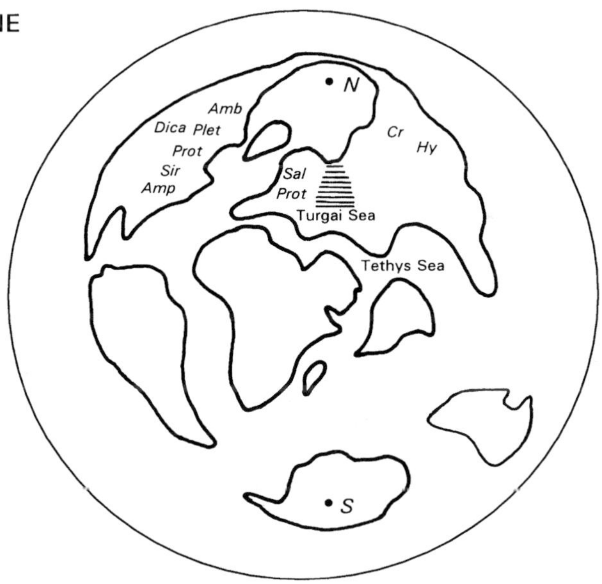

are joined to form the supercontinent, Pangaea; Upper Jurassic: the Turgai Sea divides the early salamanders into Asiatic and Euro-American stocks; Upper Cretaceous: the Mid-Continental Sea bisects North America, but the Bering Land Bridge allows the ancestral cryptobranchids to colonise North America from Asia. Palaeocene: Turgai Sea starts to withdraw but the opening of the Atlantic Ocean isolates the ancestral ambystomatids and salamandrids in North America and Europe, respectively. Based on the model proposed by Milner (1983).

into their present-day forms has occurred since the splitting of North America and Europe. However, there are a few species of newt belonging to the family Salamandridae which do occur in North America. These include the Red-spotted Newt (*Notophthalmus viridescens*) – once included in the genus *Triturus* – and the Rough-skinned Newt (*Taricha granulosa*). The ancestors of these species probably reached North America across the Bering land bridge, as the salamandrids of Europe started to diversify.

The present-day distribution of salamanders can therefore be related to continental drift and climatic changes which have occurred since the Jurassic period. The salamanders originated and radiated in Laurasia, and the continents which once formed this land mass remain their stronghold. Arid zones in southern Laurasia prevented the colonisation of Gondwana before it split from Laurasia, and this explains the absence, or near-absence, of salamanders from continents such as Australia and Africa. Two European genera, *Salamandra* and *Pleurodeles*, managed to colonise the north-west tip of Africa when this continent was connected to the Iberian peninsula in the Late Miocene, about 6 million years ago. The Sahara desert has provided an effective barrier to dispersal further south. Those salamanders which occur in South America are also fairly recent arrivals. The completion of the Panama isthmus between North and South America in the early Pliocene provided a pathway for the plethodontids to move southwards and colonise this new area.

Speciation and Glaciations

New species are formed after a population of animals becomes split up into a number of sub-units. The processes which cause such divisions are continents splitting, mountain ranges forming, incursions of seas, and the formation of glaciers. All of these present barriers to the free movement of individual animals between different areas. Once isolated from other members of its own kind, a group of animals may become adapted to local conditions, and over millions of years their descendants evolve into a new species.

Many of the present-day newt and salamander species evolved between 10 and 20 million years ago during the Miocene period. This was before the great ice sheets of the Pleistocene covered large areas of Europe, making many newt and salamander habitats inhospitable. So how did the species we know today evolve, and what effect did the glaciations have upon them? Once again, we must turn to geological history and continental drift to find an answer. Oosterbroek & Arntzen (1992) provide a reconstruction of events that may have led to the evolution of the different species of *Triturus*.

Between 15 and 20 million years ago southern Europe and Asia were divided into two parts by the Tethys Sea. To the west lay areas which were to later become the Iberian and Italian peninsulas; to the east lay the Balkans and Asia Minor. In the west two small species, Bosca's Newt and the Italian Newt (*Triturus boscai* and *T. italicus*), evolved around this time. Today, these species are endemic to the Iberian and Italian peninsulas, respectively. To the east, two groups of species evolved, corresponding to the large-bodied (or '*cristatus*') and small-bodied (or '*vulgaris*') groups of today. The large-bodied group in the east was further divided by a later incursion of the Tethys, which separated Asia Minor into eastern and western parts about 15 million years ago. Most of the large-bodied species were left in the Balkans or western Asia Minor, but the Banded Newt (*T. vittatus*) was isolated to the east. Two million years later, the seas receded again, enabling the large-bodied and small-bodied groups to disperse more widely throughout the Mediterranean region. This was followed by a period of mountain building which gave rise to the Alps and Pyrenees, about 10 million years ago. These mountain ranges isolated the Iberian peninsula from the central and eastern Mediterranean region. Palmate and Marbled Newts (*Triturus helveticus* and *T. marmoratus*) evolved in the west, while the large-bodied and small-bodied groups underwent further speciation in central and eastern areas (giving rise to *T. carnifex*, *T. dobrogicus*, *T. karelinii* and *T. montandoni*).

All of these species were subject to a further upheaval during the Pleistocene, with the coming of the ice ages. At least four glaciations occurred during this period, the most recent of which finally receded about 10 000 years ago. The ice sheets started to develop at northernmost latitudes and slowly spread southwards across Europe. As the ice spread, amphibians (and practically all other animal life) were forced southwards too. At their maximum extent, the only areas which were mild and moist enough to support salamanders and newts were in three areas at the southernmost extremities of Europe. These were the Iberian, Italian and Balkan peninsulas (Fig. 3.4). As the ice sheets absorbed a large amount of water, sea levels fell, and the Iberian and Balkan peninsulas had connections with Africa and the rest of Asia, respectively. However, the deep sea around the Italian peninsula meant that this area was still largely isolated. Isolation in these glacial refuges led to subspecies being formed. With the retreat of the glaciers, these subspecies migrated north and started to overlap once again with each other. The striped and spotted forms of the Fire Salamander are an example of this. Both forms moved back into central Europe from their glacial refuges, and today have a wide zone of overlap in central Germany. As Veith (1992) speculates, if the spotted form continues to expand westwards, and the striped form continues to move eastwards, the two subspecies will eventually become indistinguishable. A similar

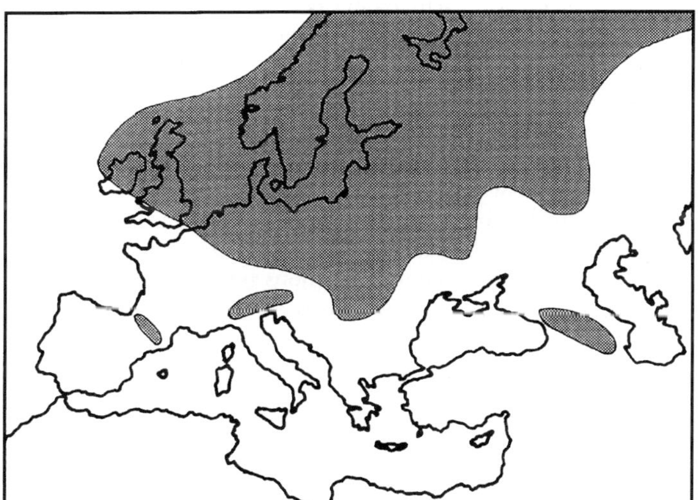

FIG 3.4 *Approximate extent of ice sheets during the Pleistocene glaciations. Forested areas in the Iberian, Italian and Balkan peninsulas provided ice-free refugia for amphibians in southern Europe. Redrawn from Steward (1969).*

pattern of overlap is seen today between newt species that were isolated in different refuges. Marbled Newts and Crested Newts, for example, expanded from their Iberian and Balkan refuges, respectively, and today occupy an area of overlap in central France where hybridisation between the species may occur (see Chapter 6).

The Pyrenees and the Alps provided an effective barrier to the northward migrations of some forms, which consequently remained in their southern refuges and became adapted to local conditions. The Golden-striped Salamander (*Chioglossa lusitanica*) and the Ribbed Salamander (*Pleurodeles waltl*), for example, are today confined to the Iberian peninsula (although some Ribbed Salamanders moved southwards into Africa when sea levels fell). Similarly, the Spectacled Salamander (*Salamandrina terdigitata*) has remained within present-day Italy.

Some salamanders were less adaptable and did not survive the ice ages. Those belonging to the family Proteidae were almost entirely wiped out. The same fate may have befallen the Plethodontidae, although evidence for a more widespread distribution of lungless salamanders in Europe has yet to be forthcoming from the fossil record. It is perhaps significant that the only surviving members of these families in Europe – the Blind Cave Salamander (*Proteus anguinus*) and the seven species of cave salamanders (genus *Hydromantes*) – are subterranean forms, and may have had ancestors who managed to survive the glacial periods by staying underground.

Sket & Arntzen (1994) propose an alternative theory to explain the current restricted distribution of the Blind Cave Salamander. Before the Pliocene, the streams that now form the subterranean habitat for this species still ran at the surface and had not yet carved their present-day underground passages. At this time the genus *Proteus* may have consisted of a more diverse group of species living in a range of surface streams and lakes. As the streams disappeared underground, a new subterranean habitat became available to these unusual salamanders. Through adaptation to living in caves, *Proteus* lost its diversity and converged to form a single, highly specialised species – *Proteus anguinus*. Sket & Arntzen (1994) further point out that there is no evidence to suggest that Pleistocene glaciations affected *Proteus* before one million years ago, by which time the modern species had already evolved. Although the pressures that led to the decline of the supposed surface-dwelling *Proteus* are unclear (perhaps competition or predation from fish resulted in their extinction), this theory suggests that the relict distribution of *Proteus* is due to the unique hydrological conditions of the region it inhabits.

Those tailed amphibians which became adapted to cool conditions at high altitudes survived the periods between glaciations in mountainous areas. The three species of brook salamanders (genus *Euproctus*) were probably more widely distributed before the ice ages. Rising sea levels isolated them in upland areas of Sardinia, Corsica and the Pyrenees where they evolved into separate species. The two species of Alpine salamander, *Salamandra lanzai* and *S. atra*, arose as a result of isolation in the south-western Alps and Balkan peninsula, respectively (Nascetti *et al.* 1988). Today, Alpine Salamanders replace Fire Salamanders at altitudes above 900 m. One hynobiid salamander reached Europe from the east after the last ice age. This was the Siberian Salamander (*Salamandrella keyserlingii*) which is known from several localities in Russia.

The Mediterranean islands of Corsica and Sardinia, together with their associated salamander faunas, separated from the land mass of Iberia about 29 million years ago. Some of the ancestral salamanders would later give rise to separate species of Brook Salamanders on each of the islands. Corsica and Sardinia remained joined together until at least 9 million years ago. The evolution of the endemic Corsican and Sardinian brook salamanders therefore dates from this time (Caccone *et al.* 1994). The four species of lungless salamanders found on Sardinia probably stem from two separate invasions. *Hydromantes genei* was the earliest coloniser, and became established on the western part of Sardinia when the island separated from the continent during the Miocene. The three other Sardinian *Hydromantes* occur on the eastern part of the island, and are actually more closely related to those in Italy than to *H. genei* on the same island. Their arrival in Sardinia is therefore probably much more recent, and may have occurred around 5.5 million years ago when a connection between Sardinia and the mainland was briefly restored when sea levels fell (Nardi 1991).

Relationships Between the Living Tailed Amphibians

About two-thirds of all the tailed amphibians that occur in Europe belong to a single family, the Salamandridae. Within this family, 12 species are 'newts', and belong to a single genus, *Triturus*. Moreover, *Triturus* represents the most widely distributed group of tailed amphibians within Europe and Asia, almost reaching the Arctic circle within Scandinavia yet extending as far south as Israel. We shall therefore take a closer look at the evolutionary relationships between members of the family Salamandridae, and in particular between those belonging to the genus *Triturus*.

The family Salamandridae

Reconstructions of the phylogeny of the salamandrids have recently been attempted by Scholz (1993) and Titus & Larson (1995). These authors used different methods in their analyses and so the evolutionary trees they produced are slightly different. However, some general conclusions can be drawn (Fig. 3.5). The most primitive members of this family include the genera *Pleurodeles* and *Tylototriton*, and Scholz suggests that the ancestor of recent forms may have resembled one of these. *Salamandrina* is also considered to have primitive characteristics, and may have diverged early in the evolution of the Salamandridae. Indeed, *Salamandrina*, *Pleurodeles* and *Tylototriton* were probably widespread across Europe during the Tertiary era. Since that time a cooling of the climate has led to *Tylototriton* retreating to eastern Asia, and *Pleurodeles* and *Salamandrina* having confined distributions in the Iberian and Italian peninsulas, respectively.

Another group of salamandrids evolved towards a more advanced terrestrial way of life. The skeleton became lighter, and the jaw structure better adapted for capturing prey on land. Males captured their mates from below, and a mechanism for transferring his spermatophore on land was developed. Some species gave birth to well-developed larvae rather than laying eggs in water, and some even became viviparous and produced fully metamorphosed young. This group comprises the 'true' salamanders: *Salamandra*, *Mertensiella* and *Chioglossa*.

The other large group of salamandrids probably evolved from a pool-dwelling ancestor, and breed in ponds, lakes and rivers. They have dispensed with the ventral mating embrace, and there is minimal physical contact between the sexes during courtship. This group comprises the newts of Europe (genus *Triturus*) as well as those of Asia (*Neurergus*, *Cynops*,

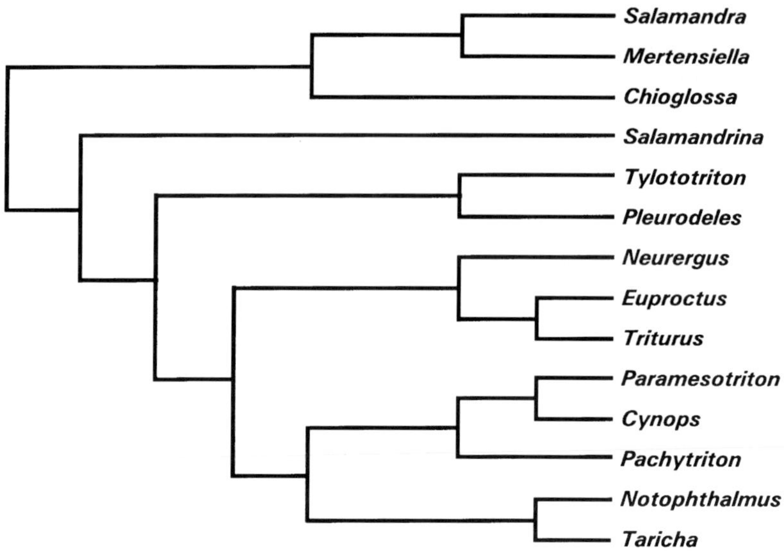

FIG 3.5 *Hypothetical evolutionary tree showing the relationships between the different genera of the family Salamandridae, based on the DNA analyses of Titus & Larson (1995).*

Paramesotriton, Pachytriton). The North American genus, *Notophthalmus,* shares some features with the Old World newts. It therefore seems likely that the ancestors of *Notophthalmus* migrated across the Bering land bridge at a later date than those of *Taricha,* another North American salamandrid which retains more primitive features.

The brook salamanders (genus *Euproctus*) lie closer to the aquatic newts than the terrestrial salamanders, but have evolved an unusual mating embrace using the tail, and have restricted distributions in mountainous regions.

The genus Triturus

The evolutionary relationships of the European newts have been subject to intense investigation in recent years. The earliest investigation was by Bolkay (1928) who classified the species according to the bones in the skull. Three groups were recognised: the 'large-bodied' newts (two species, one of which, the Crested Newt, *T. cristatus,* has now been split into four species), the 'small-bodied' newts (five species) and an intermediate 'group' which comprised just the Alpine Newt (*T. alpestris*). More recent studies have integrated morphological, behavioural and genetic information into the evolutionary picture (Arntzen & Sparreboom 1989, Macgregor *et al.* 1990). These new methods have generally confirmed the division into 'large-bodied' and 'small-bodied' species, but two slight problems remain. These concern (1) the position of *T. italicus* and *T. boscai,* in relation to the other small-bodied species; and (2) the position of the intermediate-sized *T. alpestris* and *T. vittatus* in relation to the 'large' and 'small' newts, respectively. Different methods place these species in different positions within the evolutionary tree. One way around this problem is to

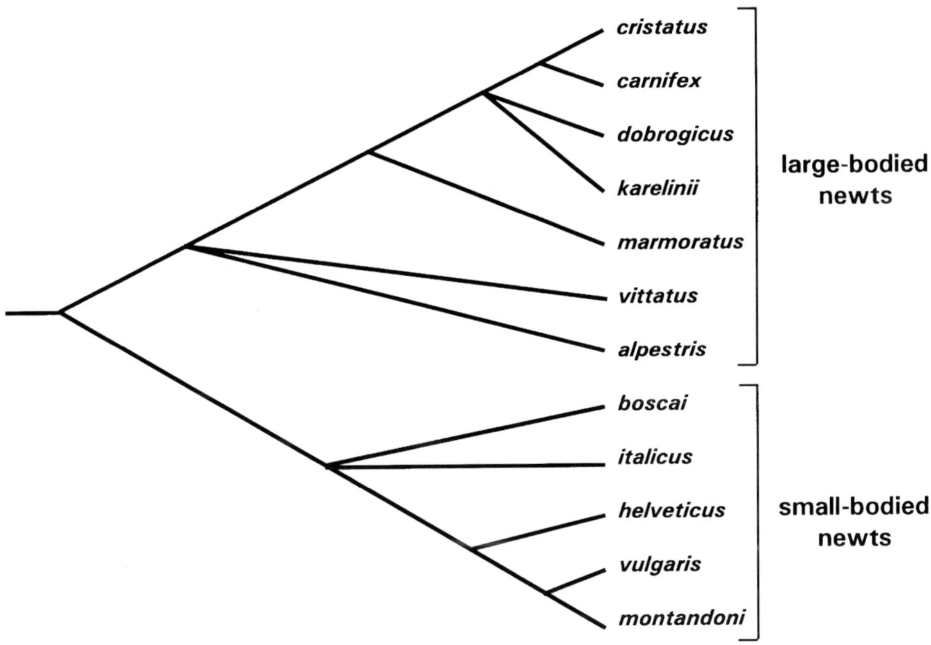

FIG 3.6 *Hypothetical evolutionary tree showing the relationships between newts of the genus* Triturus, *based on the analyses of morphology, genetics, palaeontology and behaviour by Macgregor* et al. *(1990).*

combine all morphological, behavioural and genetic data to produce a single evolutionary tree based on all the data. Such analyses have shown that *T. italicus* and *T. boscai* are actually more closely related to each other than they are to the other small-bodied newts (i.e. *T. vulgaris*, *T. montandoni* and *T. helveticus*), and that *T. vittatus* clearly falls into the large-bodied category (Giacoma & Balletto 1988, Macgregor *et al.* 1990). The most comprehensive analysis places *T. alpestris* species closer to the large-bodied newts (Macgregor *et al.* 1990). These authors therefore propose that the genus *Triturus* is divided into just two groups, with *T. alpestris* allocated to the large-bodied newt category (Fig. 3.6).

When interpreting the evolutionary relationships between different newt and salamander species, it is important to bear in mind that some characters can evolve more than once. According to Titus & Larson's (1995) analyses, for example, the 'small-bodied' and 'large-bodied' newts may represent two distinct groups which evolved quite separately from different ancestors.

CHAPTER 4

Reproduction

Euproctus asper ♂ × ♀ *mating*

ALL amphibians, whether they live in temperate areas or the tropics, go through regular cycles of reproductive activity. These cycles ensure that mating, egg-laying and larval development will take place at those times of year when environmental conditions are most favourable for these activities. These periods are usually related to the filling of water bodies, which will subsequently develop a rich community of food organisms for amphibian larvae. At the same time temperatures will be at a level that will encourage mating between adults, and allow the normal development of eggs and larvae. In tropical areas, such conditions may prevail for much of the year and breeding may be almost continuous. In temperate areas, such as Europe, however, reproductive cycles are distinctly seasonal, and geared to those periods that are generally wet but neither too hot nor too cold. However, even within Europe there are variations in climate that will dictate when breeding will take place. In northern Europe, for example, rain falls for most of the year but cold winters interrupt the annual cycle of activity. In southern Europe temperatures are milder, and are unlikely to fall below freezing in the winter. Summers are also hotter and dryer, however, so breeding activity takes place during the spring, autumn or winter months.

The only temperate amphibians which do not display clear seasonal cycles of reproduction are those that live in caves. These environments fluctuate much less in temperature than surface habitats, and the Blind Cave Salamander (*Proteus anguinus*), which is confined to subterranean waters, displays a non-cyclic pattern of breeding (Miller & Robbins 1954, Durand & Bouillon 1964). The Pyrenean Brook Salamander (*Euproctus asper*) also occasionally breeds in mountain caves, but appears to retain a seasonal cycle related to the seasons at the surface (Clergue-Gazeau 1965). Most newts and salamanders therefore display annual breeding cycles geared to the time of year that will optimise breeding success.

Reproduction uses up a lot of energy in adult amphibians. Male newts and salamanders must expend considerable periods of activity competing with others for mates, and courting females with elaborate displays. Although females generally take a more passive role in mating behaviour, they too must divert considerable resources into maturing a clutch of eggs. Accruing sufficient energy resources (in the form of fat bodies) will take some time in males and females, and yet both sexes must be ready for reproduction as soon as environmental conditions become favourable. For this reason, all newts and salamanders go through an annual cycle of physiological changes, which sees the growth and regression of gonads and hormonal changes which regulate reproductive behaviour.

REPRODUCTIVE CYCLES

The reproductive organs – testes in males and ovaries in females – not only produce the sperm and eggs (i.e. gametes) for reproduction, they also secrete the chemical messengers (hormones) which regulate breeding behaviour. Indeed, reproductive cycles are classified into two main categories, depending upon when the eggs and sperm are developed. In the *associated pattern* the reproductive organs develop immediately prior to breeding (Crews 1987). This occurs in those species that have an active period before breeding when they can feed and rapidly accumulate the energy resources required for the breeding period ahead. However, in many parts of Europe the optimal breeding period (spring) is preceded by a period of extreme cold during which amphibians become inactive and unable to feed. In these species, then, development of the testes and ovaries must occur before the onset of winter dormancy. Consequently, there is a phase of feeding and gonad development immediately after breeding has finished, which will provide the eggs and sperm to be used for reproduction the following year. This type of cycle, in which the growth of the reproductive organs is separated from the actual breeding period, is termed the *dissociated pattern* (Crews 1987), and is the one found in most European newts and salamanders. To some extent, then, reproductive success in any one year will depend upon how good feeding conditions were the year before.

In some species which live in harsh environments, there may not be enough food to develop a clutch of eggs every year, and breeding will occur every second year (biennial) or even less frequently. Montane populations of Alpine Newts (*Triturus alpestris*) and Fire Salamanders (*Salamandra salamandra*), for example, show a biennial pattern of breeding (Joly 1961, Vilter & Vilter 1963), while Alpine Salamanders (*Salamandra atra*) produce young only every 2–4 years.

The physiological changes associated with reproductive cycles are now well documented for a number of newts and salamanders. Smooth Newts (*Triturus vulgaris*) in northern Europe display a typical dissociated cycle (Fig. 4.1). The testes and ovaries reach their largest size in the autumn, when mating behaviour has finished and the newts have had enough time to feed and replenish their energy resources. In males sperm maturation

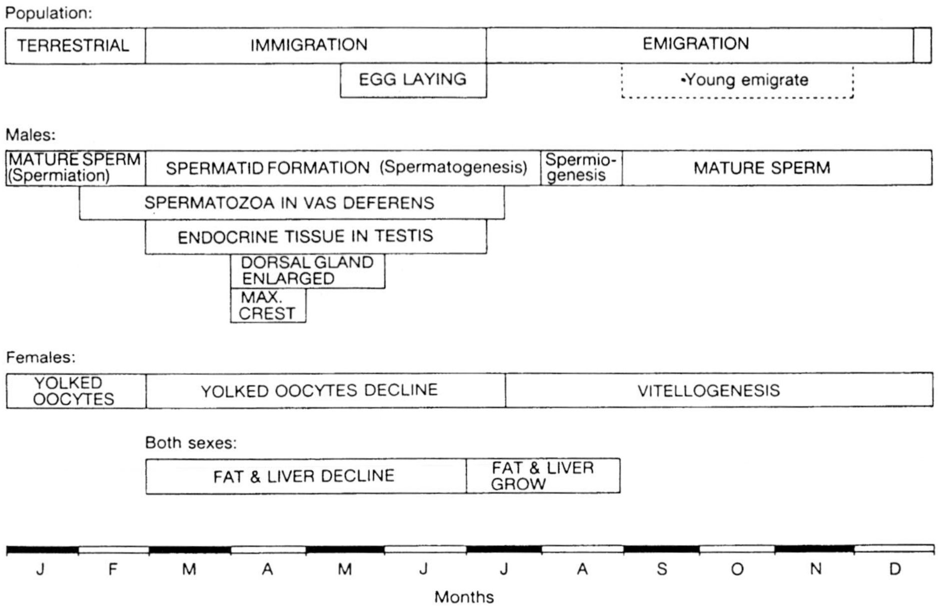

FIG 4.1 *The annual reproductive cycle of the Smooth Newt. From Verrell* et al. *(1986).*

within the testes also takes place at this time. By the time the male newts emerge from hibernation the following year the sperm has passed out of the testes (which are now much smaller) into the vas deferens, ready for reproductive behaviour in the weeks to come (Verrell *et al.* 1986). Consequently male newts have a finite supply of sperm at the beginning of the breeding season which cannot be replenished as it is used up. Similarly female newts are yolking up eggs from August onward for deposition the following year. As in males, female newts have a limited number of eggs available for laying in any one year, and the very smallest eggs will remain unyolked and be held over for a further year.

There are some variations on this cycle between species and between different parts of Europe. In north-west Italy, for example, the Alpine Newt shows a similar pattern of reproductive development to newts in northern Europe (Andreone & Dore 1992). However, reproductive behaviour starts in the autumn, and continues in the spring after a period of inactivity from December to February. There are consequently two periods of egg-laying (autumn and spring), which give rise to two cohorts of newt larvae, one of which will overwinter. This type of strategy seems to be adapted to dry summer months and wet autumnal months, which permit reproductive activity to start earlier than further north.

In the Fire Salamander mating and larval deposition have been observed at most times of the year apart from the winter. Mating seems to reach a peak in the summer, while larval deposition peaks in the spring (Klewen 1986). Consequently there is a long gestation period which lasts over the winter hibernation period for something between 6 and 12 months. However, after mating sperm may be stored for a period in the female's spermatheca until ovulation takes place, so the eggs may not be fertilised immediately (Joly 1960).

Indeed, sperm storage may last for a year or more, and Baylis (1939) reports a female salamander giving birth two and a half years after mating!

Further south in its range in Israel mild winters enable Fire Salamanders to be active throughout the year (Degani & Mendelssohn 1982), but at its south-eastern limit the reproductive cycle of the same species is adapted to the dry summers and wet winters characteristic of the region. Salamanders are inactive here during the dry summer months, and breeding is confined to October–January. Females enter ponds to deposit their young in November and December, and are mated by male salamanders as they leave the breeding pools (Warburg 1992). The breeding cycle in this area is therefore similar to that in northern Europe except that it is interrupted by a dry summer rather than a cold winter.

HIBERNATION, AESTIVATION AND ENVIRONMENTAL CUES FOR BREEDING

Reproductive cycles are clearly geared to local conditions. In northern Europe the annual cycle of activity is interrupted by a period of hibernation, when falling temperatures stimulate amphibians to seek refuge at the bottom of a pond or in cracks, crevices or burrows underground. In southern Europe, however, amphibians become inactive during the summer, when hot, dry summers interupt the activity cycle and force them into summer dormancy, or aestivation. So what are the cues which arouse newts and salamanders from their state of torpor and cause them to set off for their breeding grounds?

Whether they hibernate or aestivate, newts and salamanders generally resume surface activity during rainy periods when temperatures are in the range 2–15°C. Several studies have shown that Smooth Newts start moving to breeding ponds when temperatures rise to within the 0–5°C range. At a site in southern England, Smooth Newts were found under stone refuges throughout the year, but disappeared (presumably deeper underground) when temperatures fell below 2°C (Griffiths 1984; Fig. 4.2). The importance of rainfall in influencing surface movements is less clear. In mid-Wales Smooth Newt migrations seem to be related to temperature and the preceding rainfall, while in southern England temperature seems to be the important factor (Verrell & Halliday 1985). In Italy, however, the Southern Smooth Newt (*Triturus vulgaris meridionalis*) displays a rather different seasonal cycle. Here the temperatures never fall below freezing, and newts aestivate from June to September. Migration to the breeding pond therefore corresponds with the onset of rain in November, and lasts throughout the winter until March. The immigration to the pond consequently lasts rather longer than in northern Europe (5 months compared to around 3 months in Britain). Interestingly, emigration from the pond may be spread over a much shorter period than in ponds further north, and appears to be related to rising pond temperatures (Accordi *et al.* 1990).

In northern and central Europe Fire Salamanders are most active from March through to October, on damp nights when the temperature is 5°C or above (Szabó 1959, Joly & Lardry 1964, Klewen 1986). Activity seems to be most pronounced during thunderstorms which follow periods of dry weather. At the southern edge of its range where the same species aestivates during the dry summer, breeding activity commences when the rains arrive in November and December (Warburg 1992).

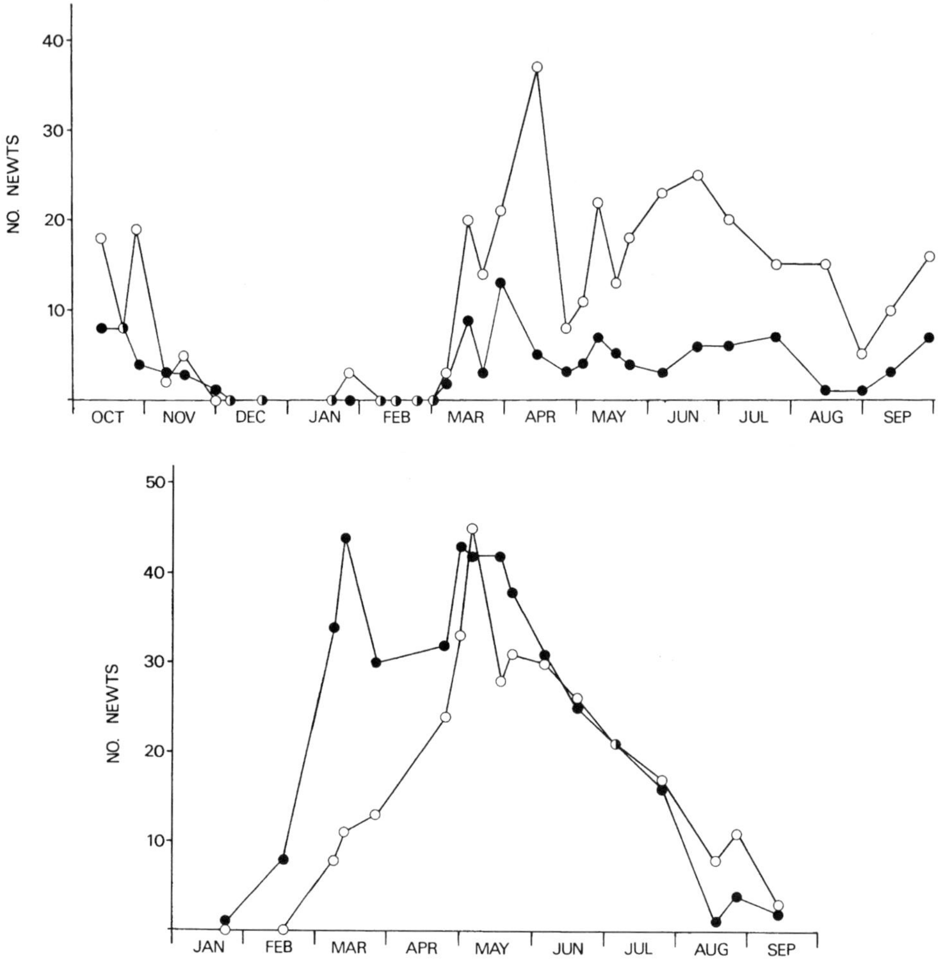

FIG 4.2 *Seasonal behaviour of Smooth Newts. The graphs show the number of newts captured under stones on land (top) and in water (bottom) around a series of ponds in North London. Solid circles, males; open circles, females. From Griffiths (1984).*

In those species which wait for ponds to fill with water each year there may be considerable year-to-year variation in the start of breeding activity, as the onset of reproduction hinges on rainfall patterns which may be unpredictable. Species which breed rather later, however, are not so dependent upon the start of the rainy season and consequently their period of reproduction may be less variable, and related more to temperature than rainfall.

With reproductive patterns so dependent upon temperature and rainfall, climate change could have important implications for the timing of breeding activities. Indeed, observations on newts in southern England suggest that these animals are already responding to warmer conditions. The dates on which Smooth, Palmate and Crested

Plate 1. (above) Spezia Cave Salamander, *Hydromantes ambrosii*. Photo F. Andreone.

Plate 2. (left) North-west Italian Cave Salamander, *Hydromantes strinatii*. Photo F. Andreone.

Plate 3. (below) Italian Cave Salamander, *Hydromantes italicus*. Photo R. Griffiths.

Plate 4. (left) Sardinian Cave Salamander, *Hydromantes genei*. Photo F. Andreone.

Plate 5. (centre left) SE Sardinian Cave Salamander, *Hydromantes imperialis*. Photo F. Andreone.

Plate 6. (above right) Supramonte Cave Salamander, *Hydromantes supramontis*. Photo T. Mutz.

Plate 7. (left) Monte Albo Cave Salamander, *Hydromantes flavus*. Photo H. Strijbosch.

Plate 8. (above) Bosca's Newt, *Triturus boscai*. Female (left) and male (right). Photo R. Griffiths.

Plate 9. (right and below) Palmate Newt, *Triturus helveticus*. A, male; B, female. Photos R. Griffiths.

Plate 10. (left and below) Italian Newt, *Triturus italicus.* A, male; B, female. Photos R. Griffiths (A) and F Andreone (B).

Plate 11. (left and below) Carpathian Newt, *Triturus montandoni.* A, male; B, female. Photos R. Griffiths.

Plate 12. Smooth Newt, *Triturus vulgaris*. A, male; B, female. Photos A. Davies.

Plate 13. (below) Alpine Newt, *Triturus alpestris*. Males (left and right) and female (centre). Photo M. Sparreboom.

Plate 14. Banded Newt, *Triturus vittatus.* A, male; B, female. Photos R. Griffiths.

Plate 15. Alpine Crested Newt, *Triturus carnifex.* A, male; B, female. Photos F. Andreone.

Plate 16. Northern Crested Newt, *Triturus cristatus.* A, male; B, female. Photos R. Griffiths.

Plate 17. Danube Crested Newt, *Triturus dobrogicus.* Male (top) and female (bottom). Photos M. Sparreboom.

Plate 18. Southern Crested Newt, *Triturus karelinii*. A, male; B, female. Photos M. Sparreboom.

Plate 19. Marbled Newt, *Triturus marmoratus*. Male (right) and female (left). Photos M. Sparreboom.

Plate 20. Pyrenean Brook Salamander, *Euproctus asper*. Photo M. Garcia-Paris.

Plate 21. Corsican Brook Salamander, *Euproctus montanus*. Photo M. Vences.

Plate 22. Sardinian Brook Salamander, *Euproctus platycephalus*. Photo H. Strijbosch.

Plate 23. Spanish Ribbed Salamander, *Pleurodeles waltl*. Photo M. Garcia-Paris.

Plate 24. A, Spectacled Salamander, *Salamandrina terdigitata*. B, belly and throat coloration.
Photos F. Andreone (A) and R. Griffiths (B).

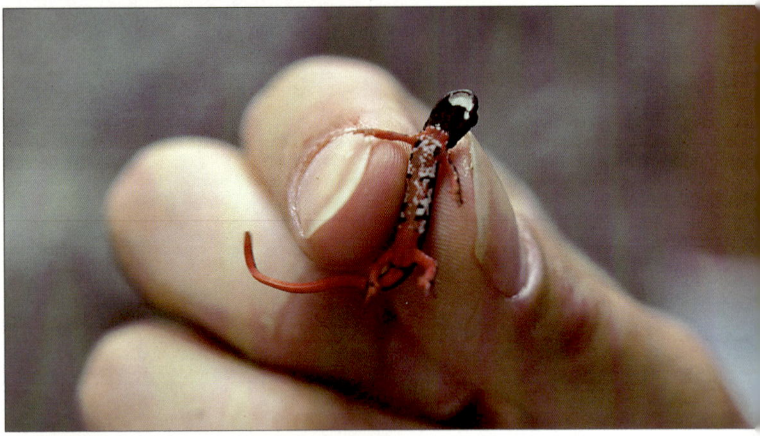

Plate 25. Alpine Salamander, *Salamandra atra*.
A, *S. a. atra*; B, *S. a. aurorae*.
Photos F. Andreone.

Plate 26. Lanza's Alpine Salamander, *Salamandra lanzai.* Photo M. Vences.

Plate 27. Fire Salamander, *Salamandra salamandra.* Photo R. Griffiths.

Plate 28. Caucasian Salamander, *Mertensiella caucasica.* Photo R. Griffiths.

Plate 29. (left) Luschan's Salamander, *Mertensiella luschani*. Photo M. Franzen.

Plate 30. (below) Golden-striped Salamander, *Chioglossa lusitanica*. Photo M. Vences.

Plate 31. (left) Blind Cave Salamander, *Proteus anguinus*. Photo C. Mlinar.

Plate 32. (below) Siberian Salamander, *Salamandrella keyserlingii*. Photo B. Thiesmeier.

Newts (*Triturus vulgaris, T. helveticus* and *T. cristatus*) arrived at a breeding pond were monitored every year from 1977 to 1994. Over this period, the timing of the first arrivals became progressively earlier by 5–7 weeks, shifting steadily from February to December (Beebee 1995). Although the populations as a whole did not display such a clear shift, the earlier breeding of the vanguard animals is closely related to temperature, which has increased over the 17 year period by about 0.2°C per year.

Simple responses to rainfall or temperature are not the whole story, however. To some extent, the reproductive cycle is regulated by an annual physiological rhythm, which ensures that the growth and regression of gonads occurs in phase with the seasonal climatic cycle. Evidence for this is seen when amphibians are moved from one geographical region to another. They do not automatically adjust to the seasons at the new location; indeed, in keeping with their old seasonal regime, they may attempt to breed at a completely inappropriate time of the year. So how can newts and salamanders maintain a physiological cycle regulated by a biological clock, with the need to respond to environmental factors, such as temperature and rainfall, which may be unpredictable? One theory suggests that amphibians may remain insensitive to temperature for part of their hibernation or aestivation period, and the biological clock actually 'switches on' a period of temperature sensitivity. Any given species, or population, must then experience so many days at or above a particular temperature before it will emerge to breed (Beattie 1985).

Orientation and Breeding Migrations

Some species live in, or around, their breeding pond for most of their lives, while others disperse considerable distances over land after leaving the pond, and therefore undergo a breeding migration to and from their breeding site each year. Certainly the powers of navigation of some amphibians rival those of many birds. Victor Twitty, a North American herpetologist who pioneered much of the research into amphibian navigation, found that Rough-skinned Newts (*Taricha rivularis*) could find their way back to their home stream even when displaced up to five miles across a mountain range! How amphibians manage to find their way so accurately has fascinated biologists for many years, but most work has focused on North American species. What is emerging is that amphibians may actually be able to use a range of cues to navigate.

One reliable cue is the position of the sun. If amphibians can learn where their breeding pond is in relation to the sun, then by maintaining a certain angle of orientation in relation to its position they may be able to use it as a direction-finder or compass. However, as the sun changes its position in the sky during the course of the day, the angle which the amphibian maintains must be constantly adjusted to account for such movement. For example, for a salamander to keep moving due south early in the morning when the sun is in the east, it must keep the sun on its left at an angle of about 90°. By the evening, however, the sun will have moved about 180° and will be sinking in the west. To keep moving in the same direction (i.e. due south) the salamander must now keep the sun on its right at an angle of 90°. In order to use the sun as an accurate compass, then, the salamander must 'know' what time of day it is and adjust its angle of movement accordingly. Thus sun-compass orientation relies on salamanders having a reliable 24-hour biological clock.

Of course, most newts and salamanders do not migrate during the day in full sun as it is usually too warm and dry. Some migration may be possible on cloudy, overcast days, even if the sun is obscured, as amphibians can detect the plane of polarised light and thereby calculate where the sun is. However, as most newts or salamanders are active on land only at night it is unlikely that many use a sun-compass to find their way to their breeding pond. Instead they use cues which are visible in the night sky, i.e. the moon and stars. Again, by learning to move in a set direction in relation to such celestial cues, amphibians may find their way to breeding ponds. Once again, though, there is the problem that the moon and stars change their positions over time. The moon, for example, passes through a seasonal cycle, during which it rises slightly later each night. This gives rise to a lunar day of 24.8 hours. Similarly, the position of stars in the sky changes seasonally. Amphibians which use lunar or stellar cues as compasses must therefore have a lunar or stellar biological clock, which allows them to adjust their angle of movement according to the time of night and time of the year.

Celestial cues are certainly not the only cues that newts and salamanders may use to find their way to and from breeding sites. Recent research on the North American Red-spotted Newt (*Notophthalmus viridescens*) and the European Common Toad (*Bufo bufo*) has found that orientation in these species can be shifted by artificial magnetic fields. As the earth's magnetic field varies from place to place, it is likely that these, and probably other species, may use it as a further navigation cue. Some very early work on the Common Frog (*Rana temporaria*) by Maxwell Savage demonstrated that this species can home in on its breeding pond by the unique odour of the algal community contained therein. More recent research on the Alpine Newt suggests that this too may be able to find its way by using the olfactory cues produced by its home pond. Joly & Miaud (1993) monitored the migrations of Alpine Newts to a small breeding pond in France. When the pond was completely covered by a tarpaulin which 'sealed in' the pond odours, they found that the newts migrating towards the pond became disorientated. Further laboratory experiments on the same species showed that newts oriented towards water taken from their home pond in preference to water from other sites. This backs up the idea that olfactory cues play a role in Alpine Newt orientation.

In terrestrial salamanders, visual cues rather than the sense of smell may be used in homing. Fire Salamanders maintain home ranges, and will return to the same hiding place after wandering up to 30 m away at night. When Fire Salamanders were moved and released away from their natural hiding places, they could always find their way back providing they were released in a familiar area. Further tests in the laboratory showed that they could learn to use local landmarks, as well as a moon-compass, to find their way around at night. Total darkness caused disorientation, and the sense of smell seems to play a minor role in Fire Salamander orientation (Plasa 1979, Himstedt 1994).

In nature newts and salamanders probably use a combination of cues to find their way to and from breeding ponds. If migrating from a long distance, for example, celestial cues may be used initially. On a dark, overcast night, the earth's magnetic field may play a more important role. As the migrating amphibian gets closer to its breeding pond, it may use pond odour and local landmarks to locate it more precisely. Clearly, as well as different species using different cues, the same species may use a variety of cues according to the time of year or prevailing environmental conditions.

The remarkable navigation abilities of amphibians has lead to the widespread belief that frogs, toads, newts and salamanders always return to the pond where they were spawned.

Certainly, many individuals will return to the same pond year after year, even if the pond dries up. However this observation does not tie in with the idea of amphibians as pioneering species able to rapidly colonise newly created but often ephemeral bodies of water. This apparent inconsistency is probably resolved by the fact that young amphibians about to breed for the first time may disperse fairly randomly and use the first water body they find (which may or may not be the one they were spawned in). Having fixated on this water body, they then return to it in following years. Thus site fidelity is established once a suitable breeding site is found.

REPRODUCTIVE BEHAVIOUR

Reproductive behaviour in newts and salamanders is very different from that observed in frogs and toads. In the anurans fertilisation of the eggs is external, and the male sheds his sperm over the egg mass as it is extruded by the female. In the vast majority of newts and salamanders fertilisation of the eggs occurs in the female's reproductive tract. The male deposits his sperm in a single package, or spermatophore, and the female must be manipulated in a way that ensures this is taken up into her cloaca where it is stored in a receptacle, or spermatheca, until the eggs are ready to be fertilised. Unlike the anurans, then, mating and egg-laying (or oviposition) occur at different times, and the male is unlikely to be in the vicinity when the time comes for the eggs to be deposited. This may be advantageous in that it allows the female to place the eggs in well-hidden places over a period of time. Secreting the eggs in this way may reduce the likelihood of them being eaten by a predator before hatching, and disperses the resulting larvae so that competition for food is reduced.

The spermatophore is a delicate structure, comprising a sperm-filled cap sitting on an elaborately shaped base. However, depositing a spermatophore for a female to pick up is an inherently unreliable way of mating. Indeed, in many species the rate of detection and pick-up by females is actually quite low (Halliday 1990). For fertilisation to be successful, the male must encourage the female to stay close to him and once the spermatophore is deposited, ensure that his mate is manipulated in such a way as to pick up the sperm cap. In some species the female remains completely passive during this procedure and is physically manoeuvred by her mate. Alternatively, a combination of visual, chemical and olfactory signals are used to accomplish this delicate procedure, and they are combined into elaborate courtship displays which are characteristic for each species.

Courtship with a mating embrace

Some salamanders manipulate the female physically by holding her in a mating embrace or 'amplexus'. The male salamander may take hold of his partner either from above (dorsal capture) or from below (ventral capture). The grip used is characterisic for each species. With the exception of the brook salamanders (genus *Euproctus*), which adopt a curious embrace using the tail, all European salamandrids take hold of the female from below. This pattern is basically the same in the genera *Salamandra, Mertensiella, Chioglossa* and *Pleurodeles.* In Luschan's Salamander (*Mertensiella luschani*) and the three *Salamandra* species, mating takes place on land; in the Ribbed Salamander (*Pleurodeles waltl*) it takes place in water, while in the Golden-striped (*Chioglossa lusitanica*) and Caucasian

Salamanders (*Mertensiella caucasica*) mating can occur on land or in water. The courtship in these genera has been described by Himstedt (1965), Joly (1966), Häfeli (1971), Arnold (1977, 1987) and Schultschik (1994), upon which the following account is largely based.

The courtship starts by the male pursuing a female, and nudging her with his nose. In *Salamandra*, the male may straddle his partner's back for a period, rubbing her with glands on his chin. In Alpine Salamanders (*Salamandra atra*) chin-rubbing is particularly vigorous and normally focused on the female's head. After these preliminary overtures, the male slides under the female's body, and locks his front legs over her front legs, so that his mate is carried in 'piggy-back' fashion (Fig. 4.3). Once in this position the body and tail of the male salamander move from side to side, so that his back and hindquarters are rubbed against the female's belly and cloaca. He may also raise his head and rub it against his

FIG 4.3 *Courtship postures in newts and salamanders. A, without amplexus (e.g. newts); B, ventral amplexus (e.g. Fire Salamander); C, tail amplexus (e.g. Brook Salamander). Stippled animal is the male in each pair. P. Benson.*

partner's throat. Again, the female is stimulated by secretions from glands on the male's head during this procedure. The Golden-striped Salamander displays two additional movements which seem to be absent in the genus *Salamandra* – a vigorous side-to-side sweep of the anterior half of the body, and a curious twisting movement, where the coupled salamanders roll over several times with their tails entwined (the latter has also been observed in *Mertensiella*). In *Mertensiella* the male salamander has a soft, club-shaped structure at the point where the tail joins the back. This structure stimulates the female cloaca directly and may also release courtship odours (Fig. 4.4). The time salamanders spend coupled in this way varies between species, lasting from a few minutes in Fire Salamanders to several days in Ribbed Salamanders. Indeed, the latter species possesses horny pads on the front legs to help maintain the grip.

Courtship culminates with the male depositing a spermatophore on the substrate and manoeuvring the female in such a way that it adheres to her cloaca. The genera *Salamandra, Mertensiella* and *Chioglossa* do this by sliding out from under the female, while still maintaining their forelimb grip, thereby lowering her cloaca onto the spermatophore. This is a tricky manoeuvre, and sometimes the spermatophore adheres to the hind legs or is missed altogether. Arnold (1987) suggests that the fleshy spike on the back of the male *Mertensiella* acts as a spermatophore mimic, and that by inserting it into the female's cloaca the male ensures that she is in the right position to be lowered onto the spermatophore that he has just released. Ribbed Salamanders mate exclusively in water, and the male deposits his spermatophore in front of the female's nose. Maintaining the front-legged grip, and using it as

FIG 4.4 *Male Luschan's Salamander, showing the curious tubercle at the base of the tail (M. Franzen).*

a pivot, he then moves the female around by 180° so that her cloaca passes over the spermatophore.

The brook salamanders are unique among the amphibians in that courtship involves restraining the female using the tail (Fig. 4.3). This structure is somewhat longer in males, and is twisted around the female's lower abdomen after her mate has pushed his body underneath hers. In the Pyrenean Brook Salamander (*Euproctus asper*) the male manoeuvres the female so that the two cloacas are pressed against each other and the spermatophore is transferred directly. Corsican (*E. montanus*) and Sardinian (*E. platycephalus*) Brook Salamanders may strengthen their hold by also using the mouth or hind legs to grip the female. Males of these two species also use the hind legs to manipulate the spematophore into the female's cloaca.

So what are the advantages in maintaining a firm physical grip on your partner during courtship? First, by holding his mate in amplexus, the male salamander is effectively guarding her against other males. If another male comes along he will first have to displace the existing male before engaging the female in his own mating embrace. This may be very difficult. Second, by restraining her in amplexus, the male has a chance to stimulate an otherwise unreceptive female using tactile and chemical cues. Third, the mating embrace may ensure that the delicate procedure of spermatophore transfer is accomplished accurately. However, these three problems are common to nearly all salamanders and newts, and in the next section we shall discuss how they are solved in those species that do not maintain a mating embrace.

Courtship without a mating embrace

The courtship patterns seen in the European newts are amongst the best described for all amphibian species, and have been reviewed by Halliday (1977, 1990) and Arntzen & Sparreboom (1989). As there is little, if any, physical contact between the sexes during courtship, the male develops elaborate breeding characters on his body which serve to attract and stimulate potential mates. These characters (the *secondary sexual characters*) develop only after the males have entered water, and include a high crest stretching from the back on to the tail, bright colours on the tail and flanks, and sometimes flaps or webbing on the hind feet. Indeed, the European newts display one of the highest degrees of sexual dimorphism seen in any group of tailed amphibians (Fig. 4.3).

Courtship in all species involves three main phases. First there is a period called the *orientation phase*, during which the male newt tries to attract the female's attention by pursuing her, sniffing her cloaca, and taking up a position in front of her. Sometimes the female swims away, but if she is interested she may remain stationary, and turn to face the male. The male then embarks on the second, and longest part of his courtship, known as the *display phase*. It is during this phase that the male must persuade the female that he is a worthy suitor, and he employs a wide range of tail and body movements that are characteristic for different species. At least 12 different stages have been recognised as occurring during the display phase, and they are employed in different combinations in different species (Fig. 4.5; Tables 4.1 and 4.2).

All of the small newts use tail-fanning in their displays. This involves folding the tail back against the body and rapidly fanning the end so that a current of water containing odours released by the male is wafted towards the female. Some species, notably the Palmate Newt

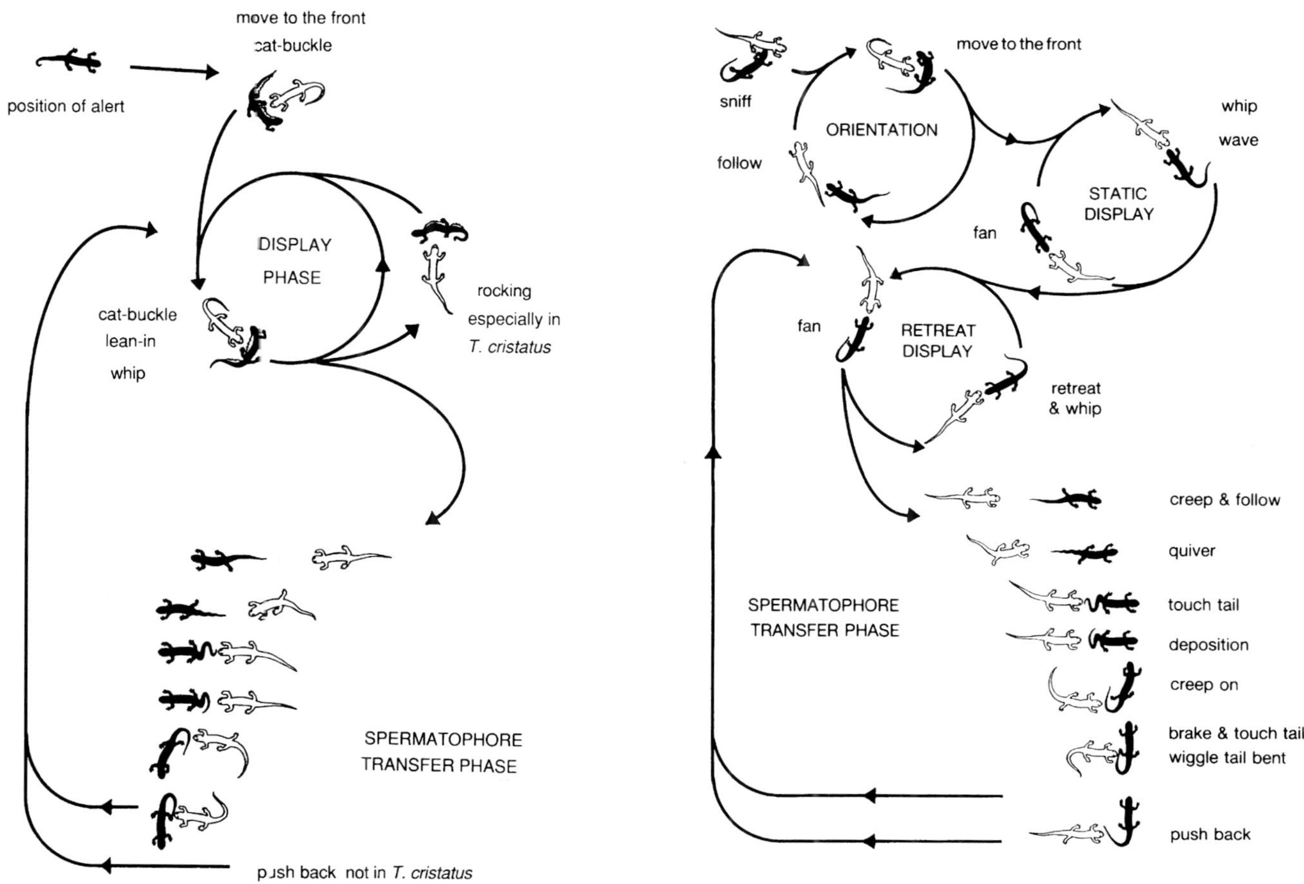

FIG 4.5 *Courtship sequence in the Smooth, Palmate and Carpathian Newts (right) and the Crested and Marbled Newts (left). Males are in black; females in white. From Arntzen & Sparreboom (1989), based on Halliday (1977).*

TABLE 4.1 *Description of behaviours performed during the display phase of courtship by male newts* (Triturus). *Based on Sparreboom & Arntzen (1987)*

Fan: Tail folded back against the flank nearest the female and end vibrated rapidly
Wave: Male presents full lateral view of his body and tail to the female
Whip: Tail lashed against flank
Whiplash: More powerful whip performed while newt is on front legs
Flick: Repetitive, rapid movement of the bent tail
Flamenco: Tail raised at 90° angle of body and tip is waved from side to side
Wiggle (distal luring): Tail tip is raised and wiggled in worm-like fashion
Cat-buckle: Body raised on hind legs and kinked in middle to emphasise height of crest
Lean-in: Leans over sideways towards the female from cat-buckle position, displaying full lateral view of body and crest
Rocking: Body rocks on forelegs while tail beats in direction of female
Head-bob: Vertical up and down movement of the head
Flurry: Rapid vibration of tail tip on side away from female performed in lean-in position
Retreat display: Female responds to male who retreats in front of her while still performing fan, whip or wave

TABLE 4.2 *Summary of the various behaviours observed during the display phase of courtship in newts* (Triturus) *(adapted from Sparreboom & Arntzen (1987); Arntzen & Sparreboom (1989))*

	Fan	*Wave*	*Whip*	*Whiplash*	*Flick*	*Flamenco*	*Wiggle*	*Retreat*	*Lean-in*	*Cat-buckle*	*Rocking*	*Head-bob*	*Flurry*
cristatus	–	–	–	+	–	–	–	–	+	+	+	–	–
marmoratus	–	–	–	+	–	–	–	–	+	+	+	–	–
vittatus	–	–	–	+	–	–	–	–	+	+	–	+	+
alpestris	+	–	–	–	–	–	–	–	+	+	–	–	–
boscai	+	–	–	–	+	+	+	–	–	–	–	–	–
italicus	+	–	–	–	+	+	+	+	–	–	–	–	–
helveticus	+	+	+	–	–	–	+	+	–	–	–	–	–
vulgaris	+	+	+	–	–	–	+	+	–	–	–	–	–
montandoni	+	+	+	–	–	–	+	+	–	–	–	–	–

(*Triturus helveticus*) and the Italian Newt (*T. italicus*), have ridges running along each side of the flank that give the body a concave shape when the fanning position is adopted. This may further assist in channelling odours towards the female. Fanning may be the main tail movement used during the display phase (Alpine, Bosca's and Italian Newts), or it may be interspersed with 'waves' and 'whips' (Smooth Newts, and sometimes in Palmate and Carpathian Newts). The wave and whip are absent in Bosca's and Italian Newts. However, these two species use two other tail movements called the 'flick' and 'flamenco' (Giacoma & Sparreboom 1987, Rafinski & Pecio 1992, Faria 1993).

During the first part of the display phase, the various tail movements are performed while the male is stationary in front of the female. If the female shows interest and starts moving towards the male, he starts moving backwards, while maintaining his display movements. This *retreat display* is seen in all the small newts, with the exception of Bosca's Newt, but the frequency of the different tail movements may change during this second part of the display phase (e.g. whips become more frequent in Smooth Newts, flicks and flamencos more frequent in Italian Newts).

The display phase of the large newts (i.e. the Crested Newt 'species group', Marbled and Banded Newts) generally lacks the delicate tail movements of the small newts, although a modified fanning display occurs in the Crested Newt (Green 1989). Courtship in these animals consists of a waving of the tail coupled with a more violent lashing of the tail over the female's head and body (the 'whiplash'). This display may be performed in a hand-stand position by raising the hind part of the body off the substratum and leaning towards the female so that the tail or flank is touching her snout. The body may be sharply kinked ('cat-buckle') during this 'lean-in' phase, and in Crested and Marbled Newts, the male newt may rock from side to side (Fig. 4.6). This position gives the female a particularly good view of the male's crest and emphasises its height. There is no retreat display in the large newts, or for that matter, in the intermediate-sized Alpine Newt.

If the male newt has been successful in maintaining the interest of his potential mate, he will proceed to the third phase of courtship, the *spermatophore transfer phase.* Up to this stage the 'large' and 'small' newts have adopted rather different patterns of behaviour, but from now on the process of spermatophore transfer is basically the same in all the *Triturus* species, as well as closely related genera. The male turns away from the female and leads her away by 'creeping' ahead of her, his tail quivering in the process. If the female is still responding she will follow and touch the male's tail with her snout. This acts as a signal to

FIG 4.6 *The cat-buckle position in a displaying male Alpine Crested Newt (M. Sparreboom).*

the male to release a spermatophore. Having deposited this on the substrate, he must now manipulate the female in a way that ensures that she picks the spermatophore up. So after moving on about one body length, he stops, turns about 90°, and blocks his mate's progress forward. The female presses forward a little further so that her snout touches her partner's tail again, and his response is to push her back slightly by unbending his tail. In doing so her hindquarters are rubbed against his spermatophore so that the sperm cap adheres to the female's cloaca.

The only species which show any deviation from the sequence of behaviours in the spermatophore transfer phase are Crested and Marbled Newts, which do not usually display the 'push back' stage.

Although the small newts and large newts show clear differences in the general pattern of courtship, Alpine Newts perform a relatively simple display which combines features of both groups, and which reflects their intermediate taxonomic position. During the display phase the only tail movements performed by Alpine Newts are tail-fanning and the wiggle (or distal lure), both characteristic of small newts (Fig. 4.7). However, it also takes up the cat-buckle and lean-in position which are only observed in the large newts, and a tail touch is not always necessary to stimulate the male to deposit a spermatophore (the first spermatophore may be deposited without a tail touch; later depositions require this stimulation from the female).

A female newt may mate with several males during the course of the breeding season, and during a sexual encounter can pick up several spermatophores from the same male. In the latter case, females are more likely to pick up the spermatophores deposited later on during

FIG 4.7 *The fanning display in the Alpine Newt (T. Halliday).*

the encounter than those deposited earlier (Halliday 1974). Thus males which are able to deposit a large number of spermatophores may be more successful than those which can only deposit a few. So far there is no evidence that female newts can store sperm from previous matings over the winter period. Any sperm retained in the female's reproductive tract at the end of the breeding season are not capable of fertilisation or are resorbed (Pecio 1992).

Despite being classified in a completely different family, the Blind Cave Salamander (*Proteus anguinus*) has a courtship display very similar to the newts. Males employ elaborate waving and fanning movements of the tail to court females, and there is no amplexus between the species. As in the newts, then, the female must be persuaded to pick up the male's spermatophore by his display. Clearly, visual cues cannot be used in this blind species which lives in total darkness. Odours and water currents generated by tail movements must therefore be the main stimulants, and this has led to a remarkable convergence of behaviour with the small-bodied newts.

Very little is known about the courtship behaviour of the lungless cave salamanders (genus *Hydromantes*). The male cave salamander has a chin gland which he may rub over the female's nostrils to stimulate her. If courtship follows the pattern observed in other lungless salamanders, there is no mating embrace as such. Instead the female straddles the male's tail and follows him in what is termed the 'tail-straddling walk'. The method of spermatophore transfer is unknown. Clusters of eggs are laid in crevices where the female salamander remains to guard the eggs. Development can take as long as a year.

Mate choice in newts

Clearly if a female is not restrained in amplexus during courtship she has the opportunity to be selective in her choice of mate. So what are the criteria that female newts use to assess the fitness of potential suitors? The one conspicuous character that male newts have in common during the breeding period is a dorsal crest, which extends from behind the head along the back and on to the tail. The crest does not start to develop until male newts enter the water, and its size is dependent upon the condition of the male on arrival at the pond. Males which have previously fed well will have accrued sufficient stored reserves to develop a large crest (Baker 1990a). Big crests may therefore indicate good feeders, and a female newt selecting a big-crested male will be ensuring that her eggs are fertilised by a male that is a highly efficient forager. Large crests may also be a general indicator of body size, and as newts continue to grow throughout life (albeit rather more slowly when older), a large crest may indicate to a female that the male is older and therefore a 'good survivor'. Large-crested males may also produce more spermatophores, a further indication that he is a fit individual who can afford to divert more resources into sperm production than a smaller crested counterpart. Despite the apparent overwhelming benefits for females in selecting males with larger crests, it may seem surprising that the evidence that they do choose males on the basis of crest size is rather sparse. Part of the problem is methodological, in that large-crested males also have a larger overall body size, may display more vigorously, and produce more courtship pheromones. All of these characters may influence a female's choice of mate, but even in a laboratory situation, teasing out the importance of each one is extremely difficult.

Courtship display can be a long and exhausting procedure for a male newt, particularly as he must periodically return to the surface and take a gulp of air. In order to maintain the interest of the female, he must obviously try to avoid surfacing for as long as possible, so his elaborate display is carried out while he is holding his breath. Males which can hold their breath for longer are therefore more likely to sustain the interest of females (Halliday & Sweatman 1976). To some extent all amphibians breathe through their skins when in water, and the development of a crest increases the surface area over which gaseous exchange may take place. Indeed, male newts with bigger crests can hold their breath for longer. A large crest therefore confers an indirect advantage in that males with such an adornment are less likely to have to break off from courtship to take a gulp of air. In general, then, females do tend to mate with males with larger crests, but they may not be actively choosing them on the basis of crest size.

A further question is how selective male newts are in choosing females. Casual observations suggest that males are fairly indiscriminate in this respect, courting any female who happens to be in the vicinity with equal intensity, sometimes even if she is of a different species! Laboratory observations have shown that males are capable of being selective, and given the choice, will pick females that are rather larger and fatter, and therefore carrying more eggs (Verrell 1986). The number of sperm in a single spermatophore outnumbers the number of eggs carried by the female, but by choosing a large mate, the male is ensuring that he is fertilising as many eggs as possible.

Territorial and non-territorial mating systems

In the Alpine Newt and all of the small newts males devote most of their energy to displaying to females and there is very little evidence of any aggressive male–male interactions. Despite the lack of aggression, however, males may interfere with each other's courtship in order to secure a mating. One intriguing form of sexual interference observed in the Smooth Newt and some other species is where interfering males become 'female mimics' during courtship. After allowing another male to do most of the courtship, an interfering male will move in and stimulate his rival to deposit a spermatophore by touching his tail in the same way that the female would. He will then proceed to lead the already receptive female away and complete the courtship with his own display (Verrell 1984). A more physical technique is adopted by the Ribbed Salamander. In this species the male may release his grip on the female during sperm transfer by unlocking one front leg. Other males may grab the loose appendage and try and displace the coupled male from his mate (Arnold 1977).

Male newts may outnumber females in a pond and sexual interference may consequently be very common. Very often a male will increase the intensity of his display if a rival is nearby, and the retreat display in small newts may serve to manoeuvre females away from potentially interfering males. Indeed, Halliday (1990) points out that sexual interference is an ancient pattern of behaviour that is present in at least three salamander families (Plethodontidae, Ambystomatidae and Salamandridae), and that we should, perhaps, regard the elaborate courtship patterns as ways of defending females against other males. What is clear is that in the small newts there do not appear to be any particular areas that are used for courtship in preference to others – males seem to wander widely within a pond, and may even wander between ponds if two or more are close together.

The large newts are rather different in that they have certain areas within a pond which are used for displaying to other newts, and these may be aggressively defended against rival males. In territorial defence the large newts display to other males in the same way that they display to females, and observations in natural ponds have shown that interactions between males are actually more frequent than male–female interactions (Zuiderwijk & Sparreboom 1986, Hedlund & Robertson 1989). This system is equivalent to the 'lek' system seen in many birds, where a group of males are spread over a restricted area, each defending an individual display area. The display areas are used by male newts only for courting females and displaying to other males. Although some newts regularly use the same part of the pond for displaying, some of the display areas are transient, and one male newt may use many different areas during the course of an evening's courtship. Aggression between males increases in the order Crested, Marbled and Banded Newts. Biting has been observed in male–male encounters in all three species, but the Banded Newt is particularly aggressive in this respect. The display areas tend to be flat, open areas which allow newts to see each other clearly and provide an even surface for spermatophore transfer. Like the Smooth Newt, Crested Newt males may interfere with another male's courtship by mimicking female behaviour. Males may adopt this behaviour even in the absence of a female newt (Zuiderwijk 1990). A male in possession of a display site may display to an approaching male using the typical lean-in and rocking behaviour of a courting male. The approaching male may then nudge and sniff the displaying territory-holder in the same way that an interested female would. In this way, the female mimic may lead the male away from his lek, and then return to take possession himself. Interestingly, such sexual interference by female mimicry occurs less frequently in the Marbled Newt. Males of this species, with their more aggressive style of defending territories, seem to have a higher degree of success in retaining females they have attracted to their display area (Zuiderwijk 1990).

Giacoma & Balletto (1988) suggest that the number of display sites available within a pond may be limited, and that strong male–male competition for them has evolved as a result. They further argue that visual cues are particularly important in the courtship of the large newts. It is in these species that the crest is the largest, and the lean-in position provides a particularly good view of the male's adornments to the female. It may be significant then, that the Banded Newt is not only the most aggressive *Triturus* species but also the one with the largest crest and most striking colouration. That Crested Newts, and possibly Marbled Newts, lack the push-back movement during spermatophore transfer phase, could be related to the fact that the delicate manipulation of the female over the spermatophore is somewhat easier in flat, open areas.

The small newts, on the other hand, are more often found in shallow, densely vegetated water, where visual communication may be difficult. Consequently their breeding adornments are rather less conspicuous than those of the large newts, and have evolved primarily to assist in the channelling of chemical odours released by the male. Italian, Bosca's, Palmate and Carpathian Newts, for example, all have small crests, but ridges along the body help create a channel for odour transmission during the fanning display, which is the main feature of courtship in these species. In the small newts, then, olfactory communication may be more important than visual communication during courtship. Smooth Newts do not quite fit this pattern (they are a small species with some visual features of a large species), but the idea is supported by the relative size of the odour-secreting dorsal gland in different species. This gland releases the odours used in courtship, and is rather larger in the small newts than in the large newts. Giacoma & Balletto (1988) also point out that territorial defence may be difficult in well-vegetated areas, and that the retreat phase of courtship may have evolved instead as a mechanism for males to lead females away from other, potentially interfering males.

Territoriality, or the lack of it, has been little studied in the other European salamanders. Kastle (1986) describes some interesting observations on male–male interactions in a population of Fire Salamanders in the Bavarian Alps. In this region male salamanders appear to defend prominent landmarks, such as stone heaps or raised tree roots, against other males. Combats between rival males may ensue, involving chasing, straddling of each other's backs, and most remarkable of all, a posture in which rivals rise on their hind legs and wrap the front legs around each other in a face-to-face wrestling match. Some males return to the same 'look-out' posts each year, and the winner of the salamander fights is usually the older individual. Similar observations have been made by Andreone (1992) on Lanza's Salamander (*Salamandra lanzai*) in the Italian Alps. As in the Fire Salamander, males emerged from their hiding places during wet weather and adopted positions on elevated landmarks which gave a good view of the surrounding area. One pair of males were also found locked in a tight embrace and rubbing heads against each other. Such behaviour could have resulted from a territorial dispute. Pyrenean Brook Salamanders also respond aggressively to other males. When a territorial male encounters an intruder, he fans his tail rapidly, bites the other male and marks his territory by rubbing secretions from his cloaca on the substrate (Lengvenus & Parzefall 1992).

EGG-LAYING OR LIVE-BEARING?

There are three means of producing young in amphibians. The most widespread strategy, *oviparity*, simply involves depositing eggs in water or a damp place. In *ovoviviparity*, the eggs are retained within the female for a period where they hatch within the oviduct. In this case the female will give birth to larvae. If the eggs are retained within the female to the extent that they exchange gases, waste products and nutrients with the maternal blood stream, then the strategy is *viviparity*. In fact there are various degrees to which the female may nourish the developing embryo within her oviduct. Even in ovoviviparity there may be exchange of gases across the egg membranes, while in advanced viviparity a placenta-like interfolding of the egg membranes and oviduct wall may occur to allow nutrients to pass from the maternal to the larval bloodstream. Some authors argue that there can be no hard and fast dividing line between ovoviviparity and viviparity, and recommend the use of the terms oviparity and viviparity to simply distinguish between egg-laying and live-bearing. Indeed, in some species the mode of development within the oviduct is largely unknown.

Fire and Alpine Salamanders provide a nice illustration of how live-bearing has evolved in response to environmental conditions. In the Fire Salamander the eggs are fertilised in the oviduct, where they are supplied with oxygen from the mother and develop to an advanced stage. The female salamander therefore gives birth to larvae which have well-grown limbs and can feed and swim independently. This pattern is taken a stage further in the Alpine Salamander, in which only one or two eggs develop out of a clutch of 20–30. As soon as the two fertilised eggs use up their yolk supply, the remaining unfertilised eggs provide food for the developing embryos. They continue to feed in this way for a year or more, and when the eggs run out the embryos start to consume cells scraped directly from the mother's reproductive tract, with the aid of specialised teeth. This method of feeding will see the embryos through to full term, and after a 2–4 year gestation period, the mother gives birth to just one or two fully developed young salamanders (Vilter & Vilter 1960, Joly *et al.* 1994). This type of pattern of development is an adaptation to the high altitudes where the Alpine Salamander is found, as in such extreme environments the growing season is just too short to allow any sort of free-living larval stage. Interestingly, some populations of Fire Salamanders from montane regions also go some way towards this pattern. These salamanders retain the larvae within the oviduct for much longer than normal, and like the Alpine Salamander, some of the embryos develop at the expense of others. Indeed, the larvae hatch within the mother's reproductive tract where they feed directly on unfertilised eggs and even other hatched larvae! Having cannibalised their siblings, fully developed salamanders will be born after a long gestation period (Alcobendas *et al.* 1994, Joly *et al.* 1994).

Like *Salamandra*, the genus *Mertensiella* displays some flexibility in its method of egg development. After an initial courtship in the water, the Caucasian Salamander (*Mertensiella caucasica*) mates on land and lays eggs in inaccessible cracks and crevices (Schultschik 1994). In Turkey its close relative, Luschan's Salamander (*Mertensiella luschani antalyana*), is more similar to the Alpine Salamander in that it is viviparous and gives birth to just two, fully developed young. However, the gestation period of around a year is somewhat shorter in Luschan's Salamander than it is in the Alpine Salamander (Özeti 1979).

So why do some species lay eggs while others give birth? One general theory for ectothermic animals is that live-bearing is favoured at more northerly latitudes and in mountainous areas where temperatures are colder and the period available for feeding and growth shorter. However, some live-bearing species, such as Luschan's Salamander, are found at relatively low latitudes and altitudes where winters are quite mild. What seems to be a broader, and rather better explanation of live-bearing is that it is associated with extreme conditions of either cold winters or dry summers, perhaps coupled with the periodic lack of surface water as a result of geological conditions. Either of these two factors will interrupt salamander activity, and if prolonged will severely limit the time available for egg or larval development in water. If the climate is too cold, development would be so slow as to almost stop completely. At the other end of the spectrum, very hot summers may result in the rapid disappearance of ponds and streams altogether. Consequently both of these pressures will favour those females which can retain their young inside their bodies for as long as possible.

Another factor which has some bearing on egg-laying or live-bearing is the predation pressure faced by eggs and larvae during their development in water. Hatchling larvae are very vulnerable to a wide range of aquatic predators, but as they grow they may become too large to eat. Newts and salamanders which habitually use predator-rich waters for breeding may therefore increase the chances of their young surviving if they release them at an advanced stage of development. Once again, however, this explanation is not entirely satisfactory as many egg-laying species live in environments which have an abundance of predators. The problem with live-bearing is that a considerable cost is involved for the female. By carrying around a bulky clutch of eggs within her body, the female salamander is increasing her own risk of being eaten, and far fewer eggs can be raised than if they were deposited and abandoned within a pond. In northern Europe, mating, egg-laying and feeding has to be fitted into a relatively short period of time, and there may simply not be enough time or resources to devote to raising eggs internally. Clearly then, there are both costs and benefits associated with egg-laying and live-bearing, and the strategy used will ultimately depend upon the relative importance of selection pressures such as egg and larval predation and local climate.

Egg Deposition

Those newts and salamanders which lay eggs deposit them singly or in small clusters. Salthe (1969) identifies three modes of oviposition. Mode I is where eggs are deposited in open static waters. This is the mode adopted by all the European newts and some of the salamanders. Mode II is when clutches of eggs are deposited in hidden nest sites in running water. The Golden-striped Salamander and the brook salamanders deposit their eggs in this way. The Corsican Brook Salamander is unusual in that the female stays close to the clutch of eggs, presumably guarding them (Alcher 1981a). Mode III, which is found only in plethodontid salamanders, including the cave salamanders of southern Europe (genus *Hydromantes*), is when eggs are laid in a damp hiding place out of water and give rise to fully developed hatchling salamanders. In general, those species which lay in static waters have a higher fecundity but smaller sized eggs than those which use running water. Moreover, those species which lay small numbers of larger eggs (i.e. Modes II and III) often show some degree of parental care. Under laboratory conditions, the Blind Cave Salamander makes a small nest and one of the parents may guard the clutch, aerating it by fanning the tail to create a current of water over the eggs. All the *Triturus* species, however, abandon their eggs after deposition.

Although egg development is highly dependent upon water temperature, there is little evidence that newts and salamanders use this as a cue in selecting egg-laying sites. What seems to be more important is the distribution and abundance of aquatic vegetation. All *Triturus* species wrap their eggs individually in the leaves of water plants close to the surface of the water. Oxygen levels are likely to be highest at the surface, and the support of the vegetation prevents the egg sinking into mud at the bottom of the pond which could inhibit respiration. Some plant species appear to be preferred over others, but if no plants are available then dead leaves or even litter such as plastic bags may be used. The wrapping of each egg into a leaf is a delicate process that the female newt accomplishes by manipulating the egg and leaf with her back legs (Fig. 4.8). Diaz-Paniagua (1989) has described this procedure in detail in the Marbled Newt. Before she starts laying, the female Marbled Newt inspects the plant by smelling the leaves. If the leaf is a suitable site for an egg, she assumes an upright position above the plant, and folds the leaf with her back legs while pulling it towards her. This preparatory phase usually lasts under a minute. Still holding the leaf between her legs, she then extrudes an egg into the fold of the leaf to which it adheres by its sticky capsule. This phase takes about 4 minutes to complete, during which time the rest of the body remains motionless above the plant.

As in courtship, egg-laying must be performed while the newt is holding its breath. Ascents to the surface for air therefore occur both before and after an egg is deposited, but sometimes deposition must be interrupted by an ascent for air. Failure to deposit an egg may also occur if the leaf was too small or if an egg had previously been laid in the same place. Interestingly, large females seem to be able to hold their breath for longer and have a lower failure rate (Diaz-Paniagua 1989). The large-bodied newts (i.e. Crested and Marbled Newts) often lay a whole series of eggs on the leaf of a floating plant such as sweet grass, giving the plant a concertina-like crimped appearance. In all newts the egg-laying period is spread over several weeks. Egg-laying rate seems to depend on temperature, and in Smooth and Palmate Newts, averages between three and seven eggs per female per day (Baker 1992, May 1993).

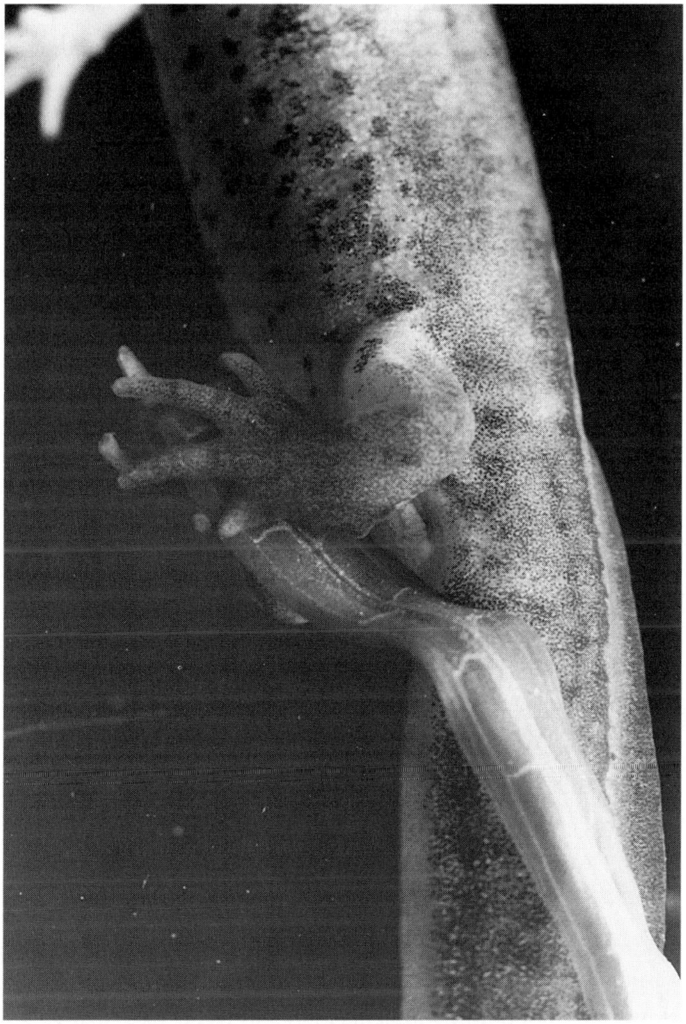

FIG 4.8 *Female Smooth Newt wrapping an egg in the leaf of a water plant (T. Halliday).*

CHAPTER 5

Egg and Larval Development

Triturus carnifex larva

WITH the exception of a few live-bearing species, all newts and salamanders deposit large clutches of eggs or larvae. This is to offset the very high mortality that both stages suffer at various stages of development. Eggs and larvae have many natural enemies in the form of predators and competitors, and may also face the risk of ponds drying out and extremes of temperature. Despite these risks, eggs and larvae have evolved structural, physiological and behavioural adaptations to help see them through this difficult period.

EGG DEVELOPMENT

The protective jelly layer surrounding the embryo actually consists of several layers of transparent capsules. Soon after deposition these capsules swell by absorbing water. In the centre of the egg the embryo itself is surrounded by a sac, known as the vitelline membrane (Fig. 5.1). In the majority of European newts the inner capsules degenerate early on in development and the embryo, still surrounded by the vitelline membrane, floats freely in

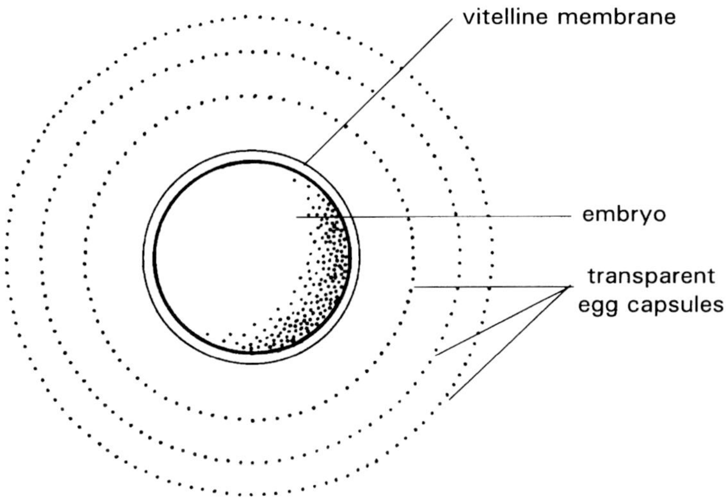

FIG 5.1 *The structure of a newt or salamander egg.*

what is called the capsular chamber (Salthe 1963). All nutrition during development is obtained from stored yolk, and waste products are passed out into the chamber surrounding the embryo. Oxygen can pass through the egg capsules and the external gills may be used as respiratory organs even before the embryo hatches.

Embryonic development in amphibians has been particularly well studied, and most basic biology texts describe the stages of larval development from the first cleavage of the fertilised egg through to metamorphosis. Such 'normal tables' for development have been produced for a wide range of species, including newts of the genus *Triturus* (e.g. Gallien & Bidaud 1959, Epperlein & Junginger 1982). Inspection of a newly laid newt or salamander egg under a hand lens or microscope will reveal one or more cleavage planes, as the embryo passes through several cycles of cell division. After a few days the embryo will have elongated, and it is possible to distinguish a rudimentary head and tail. Three pairs of external gills start to develop on either side of the head, and soon after small buds appear behind the head which will become the front legs. Earlier in development a hatching enzyme will have been released which digests the vitelline membrane, and final escape is achieved by wriggling through the degenerating egg capsules (Fig. 5.2).

An intriguing anomaly in larval development occurs in the four species of Crested Newt and in the Marbled Newt (i.e. *Triturus carnifex, T. cristatus, T. dobrogicus, T. karelinii, T. marmoratus*), and has been described in detail by Macgregor & Horner (1980). During development, 50% of all eggs of these species stop developing when the tail starts growing. The embryos spontaneously abort soon after. This curious situation is associated with the way that the structures which carry the genetic material from each parent, the *chromosomes*, come together after the egg is fertilised. There are 12 pairs of chromosomes in newt cells, and they can be distinguished under the microscope by their size, shape and pattern. The two chromosomes in pair no. 1 are different from each other, and are referred to as 1A and 1B. However only one or other of these chromosome types will pass into the individual

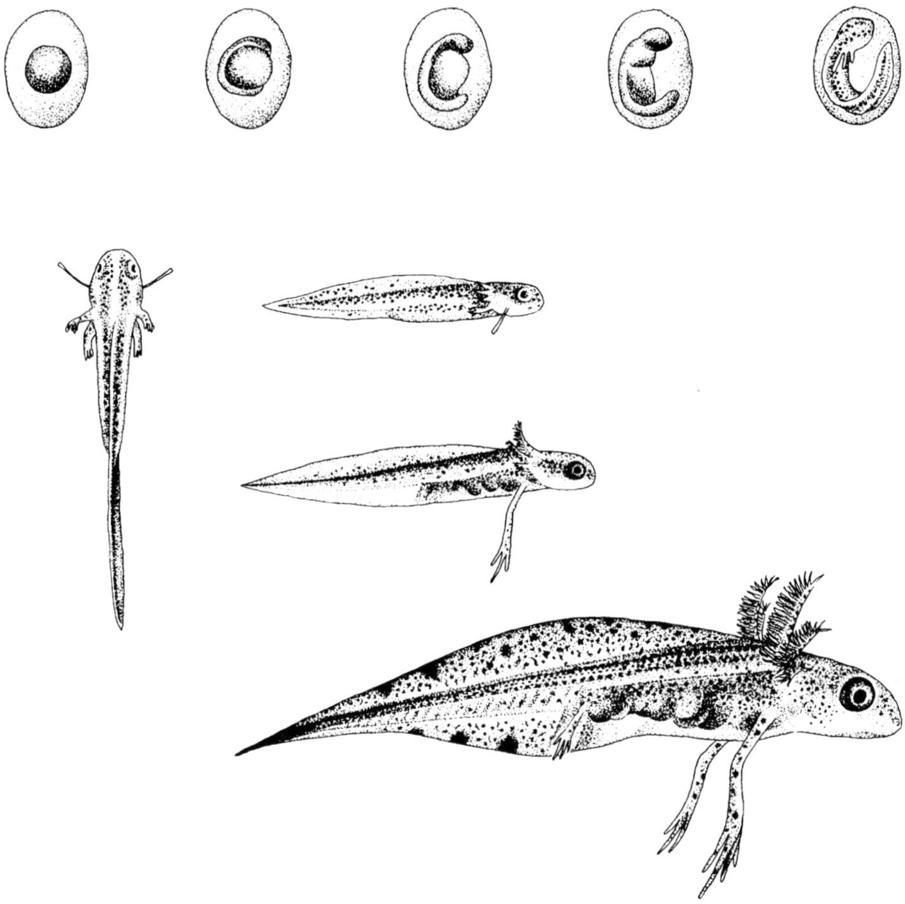

FIG 5.2 *Development of the eggs and larvae of the Crested Newt. P. Benson.*

gametes (i.e. eggs or sperm) of the adult animal. When a sperm fertilises an egg, then, two chromosomes (one from each parent) once again become paired up. If the egg receives two dissimilar chromosomes from the parents (i.e. 1A from one parent and 1B from the other), embryonic development will occur normally. If, on the other hand, the chromosomes are of the same type (i.e. 1A + 1A or 1B + 1B), development cannot proceed beyond the tail-bud stage. The end result is that half the eggs laid by Crested and Marbled Newts will always die at, or soon after, the tail-bud stage. In evolutionary terms it is difficult to explain how such a wasteful system can have evolved, and the phenonomenon remains one of nature's puzzles.

Newt and salamander eggs usually lack the dark pigment (melanin) characteristic of anuran eggs. Frog and toad eggs are often deposited in exposed areas, and the pigment may be necessary to help protect the eggs from ultraviolet radiation. As newts and salamanders normally deposit eggs in well-hidden areas such protection is not needed. However, a number of species including Smooth Newts, brook salamanders and Spectacled Salamanders (*Triturus vulgaris, Euproctus* and *Salamandrina terdigitata*) do show traces of pigment within the embryo despite being left in well-concealed sites. As Salthe (1969) points out, all amphibians which lay eggs in exposed areas seem to have protective pigment but not all those which oviposit in concealed sites lack pigment. Certainly the pigment may have other functions, such as assisting in heat retention and thereby maintaining an egg temperature slightly higher than that of the surrounding water.

The rate of egg development is influenced by a wide range of environmental factors, of which temperature is probably the most important. However, species vary in their responses to temperature, and those from warmer climates, or those that breed at warmer times of the year, can often tolerate higher temperatures than those from cooler climates or cooler seasons. Even within a species, the upper and lower limits for normal development can vary between populations and may be related to the thermal characteristics of the breeding site. Extremes of temperature at either end of the scale will kill the embryos, but within the limits for normal development, time to hatching will be much faster at the higher temperatures. In newts, for example, development at 17°C is at least twice as fast as it is at 12°C (Griffiths & de Wijer 1994; Table 5.1). Development is even faster at higher temperatures, and Crested Newt embryos develop up to six times faster at 25°C compared to 10°C (Horner & Macgregor 1985). Time to hatching can therefore be as short as a week or longer than a month, depending on temperature, although such extremes are probably rarely observed in nature.

TABLE 5.1 *Time to hatching in three species of newts. Values are mean days to hatching ± standard deviation in eggs raised at constant temperatures of 12°C and 17°C (from Griffiths & de Wijer, 1994)*

Temperature	*Crested Newt*	*Palmate Newt*	*Smooth Newt*
12°C	36.7 ± 0.22	36.5 ± 1.37	34.3 ± 1.11
17°C	15.3 ± 0.48	17.4 ± 0.55	15.6 ± 0.39

Chemical factors will also affect embryonic development but in general these have been little studied. Amphibian embryos are probably more sensitive to environmental pollutants than larval or adult stages, and even sublethal levels of such chemicals may affect the rate of development. In the vast majority of amphibians studied to date acid conditions cause hatching to be delayed. Interestingly, European newts seem to be an exception to this rule as low pH seems to result in embryos hatching prematurely and at an earlier stage of development (Griffiths *et al.* 1993).

LARVAL DEVELOPMENT

Although newt and salamander larvae are sometimes referred to as tadpoles, strictly speaking this term should be used only to refer to the larvae of anuran amphibians. Tadpoles of frogs and toads are quite different from the juvenile and adult stages, and go through a dramatic reorganisation of the body form at metamorphosis. Newts and salamanders, on the other hand, already have the basic adult body plan as larvae, and changes at metamorphosis are consequently less extreme. Hence the term 'tadpole' should be restricted to the anurans, while the more general 'larva' can be applied to either class of amphibians.

Newt and salamander larvae can be divided into two main groups – pond-dwellers and stream-dwellers (Fig. 5.3). Pond-dwelling larvae, such as those of all the newts, hatch at a relatively early stage of development. The front leg buds are still small, the mouth is not yet fully functional, and the larva remains attached to the egg capsule or adjacent vegetation by two prong-like adhesive organs attached to the head, called balancers. For a few days the larva is largely immobile, and feeds on the remaining yolk supply. By the time this is used up the mouth is functional and the balancers have been resorbed, so that the larva is ready to assume a free-swimming, predatory way of life. As the larva grows the back legs develop and locomotion is by walking on the pond bottom or by short beats of the tail. Ponds are often stagnant and low in oxygen, and large frilly gills provide a large surface area over which oxygen can be obtained. Running along the back and onto the upper and lower surface of the tail is a deep fin which also provides a large surface area for respiration. The lungs actually become functional prior to metamorphosis and larvae which live in ponds at low oxygen concentrations may swim to the surface to take gulps of air. Fire Salamander (*Salamandra salamandra*) larvae are also typical pond-dwelling forms, but as the eggs are

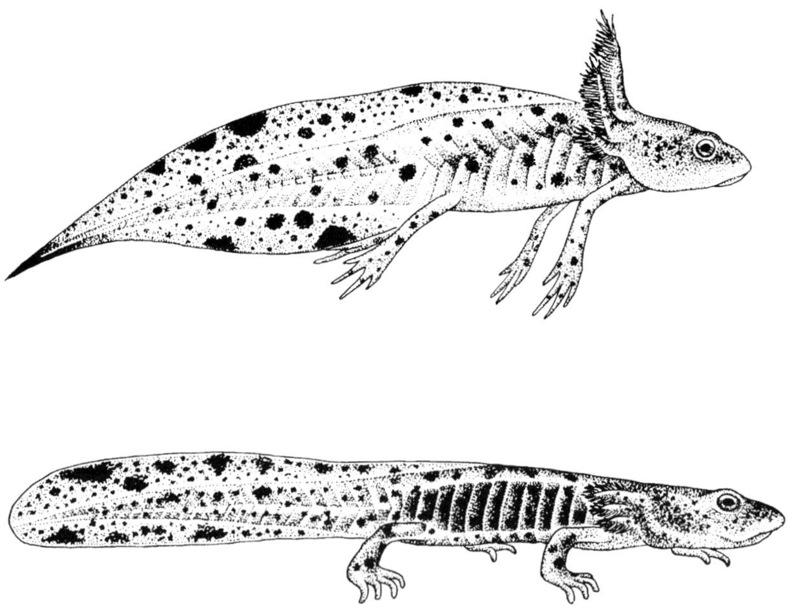

FIG 5.3 *Pond-type larva (top: Crested Newt) and stream-type larva (bottom: Sardinian Brook Salamander). P. Benson.*

retained within the female's reproductive tract during early development, the larva is deposited at an advanced stage with all four legs and a functional mouth.

Stream-dwellers, such as brook salamanders and Spectacled Salamanders, must be able to cope with flowing water on hatching, so they are more advanced on hatching with functional mouthparts and well-grown front and hind legs. Living in cool, well-oxygenated flowing water means that the gills and tail fin are less well developed than in pond-dwelling species and the lungs of stream-dwellers do not develop until the larva is about to metamorphose. Indeed, the body is generally more streamlined with a muscular tail to provide more sustained locomotion in a current of water.

Temperature has a clear effect on larval growth and development. Smooth Newt larvae raised at 20°C achieved metamorphosis in about 50 days, compared to 275 days when raised at 12°C, a six-fold difference in growth rate (Baker 1990b). Size at metamorphosis also increased over the same temperature range, but declined above 20°C. If larvae hatch late in the year when temperatures are starting to turn cooler, there may not be enough time to grow, develop and metamorphose in the same year. Such larvae may overwinter in the pond, continuing to grow over the winter months and metamorphosing the following spring. This may lead to some overlap between the larval cohorts of successive years, but overwintering larvae are usually distinguished by their larger size. Bell (1977) estimated that about 8% of the Smooth Newt larvae that survive to metamorphosis will have overwintered. This figure is greatly influenced by the nature of the pond, and in particular how long it holds water for. If the pond dries up by the end of the summer then no overwintering will occur. Alpine Newt (*Triturus alpestris*) larvae which are hatched in cool, upland ponds also frequently overwinter. If this extends over two or more winters the larvae may become neotenous (see below).

The effect of temperature on Fire Salamander larvae seems to vary across the range of this species. Independent tests on larvae from Poland and Israel have found that the optimum temperature for the development of Fire Salamander larvae from both areas was between 10°C and 20°C. Similarly, when raised at 6°C larvae from both parts of the range survived for a year or more but did not metamorphose. The main difference between salamander populations seems to be at the upper end of the temperature scale. Larvae from Poland raised at about 27°C developed normally but perished at metamorphosis. Those from Israel could withstand higher temperatures, and developed normally at up to 30°C (Degani 1982, 1984, Zakrzewski 1987).

FEEDING

Unlike frog and toad tadpoles newt and salamander larvae are exclusively carnivorous. Indeed, newts and salamanders are sometimes referred to as 'gape-limited predators' as they feed on a wide range of prey types and what they can or cannot consume seems mainly limited by prey size, or put another way, by the size of the mouth of the predator. As Smooth Newt larvae develop, for example, the number of prey items in the stomach increases, the prey items become larger, and a wider variety of prey is taken (Kuzmin & Meschersky 1987). Apart from the limitation of prey size, larvae are fairly unselective predators. This means that their diet will largely be dictated by the invertebrate community of the water body in which they live. Zooplankton, including water fleas (*Daphnia*) and copepods

(*Cyclops*), therefore frequently comprise a major proportion of their food. One study of the diet of Fire Salamander larvae showed that they consumed mainly freshwater shrimps, mosquito larvae and caddis larvae. These groups comprised some 90% of all invertebrates in the brook studied, illustrating how unselective the salamanders were as predators (Giesenberg 1991).

When two or more newt species coexist in the same pond their diets are similar, with any differences being related to size trends between the species, or to differences in habitat selection (see below). Cannibalism and predation of small larvae by large larvae may also occur. Crested Newts raised with Smooth and Palmate newts in pond enclosures had no effect on each other during early development as the three species were initially similar in size. Crested Newts are the fastest growing of the three species, however, and once they achieved a size of about 27 mm Crested Newts started predating Smooth and Palmate newts even though there was plenty of other food available (Griffiths *et al.* 1994). Certainly the larger bodied newts and salamanders seem to be more voracious predators than the smaller species, and will attempt to bite the limbs or gills of even similar-sized conspecifics if raised together.

Newt and salamander larvae have small teeth on the jaws and palate but these serve only to restrain the prey as it is swallowed whole. A combination of visual, olfactory and tactile cues are used to find prey. Indeed, the sequence of behaviours leading to prey capture is similar to that observed in adult animals (see Chapter 2). However, visual cues seem to be of overriding importance as the full range of prey-catching behaviour can be elicited using visual stimuli alone (Himstedt 1967). Food supply influences both growth and time to metamorphosis. Newt larvae raised under reduced food levels take longer to develop, and metamorphose at a smaller size (Baker 1990b). At high food levels there is a positive relationship between time to metamorphosis and size at metamorphosis, i.e. a longer period of larval development allows larvae to grow to a larger size. If food is scarce, then, taking longer to develop will not necessarily result in a larger body size (Baker 1990b).

PREDATORS

Despite the protection afforded by a jelly capsule and a folded water leaf, most eggs are consumed by predators. Bell & Lawton (1975) estimated that only 2–3% of smooth newt eggs survive to hatching. Miaud (1994) has shown that unwrapped newt eggs suffer higher mortality in the field than those wrapped in the leaf of a water plant. Further tests showed that water beetles and adult newts (particularly females) are important predators of newt eggs. Indeed, great diving beetles (*Dytiscus marginalis*) seem to be proficient at taking even those eggs that are wrapped in a leaf (Miaud 1993). Why female newts should eat their own eggs remains unclear. Other predators of newt eggs may include caddis larvae, large snails and frog tadpoles. In addition, waterfowl which eat pond weed will also consume any eggs attached to it.

Newt eggs are also prone to fungal attack, particularly from a filamentous form called *Saprolegnia*. Infected eggs are easily distinguished by their opaque, swollen appearance, and rarely hatch. Those eggs which are laid earliest, during the cooler months of the year, seem particularly prone to fungal attack. Fungal infections can spread very quickly through eggs laid adjacent to each other, and under the microscope the fungal hyphae can be seen spreading from egg to egg. Indeed, laying eggs in clusters may have been the primitive condition for tailed amphibians, with individual egg deposition evolving later as a result of predation pressure.

Newt and salamander larvae are consumed by a wide variety of predators, including larvae of larger species, and even larger individuals of the same species. Unlike some toad tadpoles which have skin toxins as a predator deterrent, newt and salamander larvae rely essentially on a cryptic way of life to avoid predation. The presence of predatory fish may go as far as excluding some species from a water body, unless there are microhabitats available where the larvae can find a refuge. Dragonfly nymphs will ambush newt larvae in weed or on the pond bottom, and both adult and larval stages of water beetles are significant predators. Given the opportunity even adult newts will prey on the larvae of their own or a different species, although their importance as predators is probably less than that of invertebrate predators.

COMPETITORS

Intraspecific competition

If amphibian larvae are crowded within a pond there may be a shortage of food, and aggressive interactions between individuals may increase. Under such conditions larvae take longer to develop, metamorphose at a smaller size, and have reduced chances of survival. Crowded conditions may also increase the variability of these factors between individuals, resulting in a wider range of sizes at metamorphosis, and a longer period of time during which newts are metamorphosing and leaving the pond. A series of laboratory experiments by Baker (1990b) has shown how two species of newts respond to intraspecific competition. Banded Newts (*Triturus vittatus*) raised at high densities were smaller than those raised at low densities, but the average larval period was unaffected. However, the variability in larval period increased with density – some larvae metamorphosed early while others metamorphosed late. Smooth Newts responded somewhat differently; those raised at high densities were the same size as those at low densities, but took longer to develop. Once again, however, larval period varied more at high densities than at low densities (Table 5.2).

TABLE 5.2 *Body size (total length) and larval period in Smooth and Banded Newts raised under high and low densities. Values are mean total length at metamorphosis and mean time to metamorphosis ± standard deviation, in larvae raised at room temperature and with* ad libitum *food (from Baker, 1990b)*

	Body size (mm)	*Larval period (days)*
Banded Newts		
High density	31.35 ± 1.11	58.35 ± 2.92
Low density	33.18 ± 0.64	57.07 ± 2.16
Smooth Newts		
High density	31.4 ± 3.52	58.1 ± 9.21
Low density	31.0 ± 2.53	46.8 ± 4.22

Intraspecific competition therefore seems to affect growth in Banded Newts and larval period in Smooth Newts, but in both species the variability in larval period is also affected. The behaviour of Banded Newt larvae under competitive conditions may therefore differ from that shown by Smooth Newts.

One explanation of how larvae can metamorphose at the same size under crowded conditions is 'competitive release'. As the largest larvae metamorphose from the population the density of the remaining individuals will fall. Any smaller individuals can then 'catch up' in terms of growth, and metamorphose at the same size, but somewhat later than counterparts that left the pond earlier. This will also mean the time scale over which the individuals metamorphose will be rather longer than under uncrowded conditions.

The densities of newt larvae used in such laboratory experiments are invariably higher than natural densities, and whether such trends in growth and development occur in natural ponds is unclear. Certainly, intraspecific competition in natural ponds is likely to hinge on whether food resources are limited or not. The abundance of invertebrate food available to newts and salamanders will show marked seasonal fluctuations as well as varying considerably between ponds. At least one field study suggests that food for newt larvae may be unlimited in natural ponds (Bell & Lawton 1975). However Baker's (1990b) laboratory experiments were carried out with newt larvae raised under *ad libitum* food. This means that the competitive affects observed cannot be attributed to food shortages, and suggests that behavioural interactions were responsible for the patterns observed.

Interspecific competition

The impact of competition from a different species is largely the same as that from intraspecific competitors. Theoretically, the effects of interspecific competition are less than those of intraspecific competition, as to some extent larvae of different species will differ in size, morphology or behaviour and thereby occupy somewhat different niches within a pond. In other words, a larval amphibian is more likely to interact with a member of its own species than with a member of a different species. However, there are exceptions to this rule. A larger, dominant competitor, for example, may have a greater impact on a smaller species than just more individuals of the smaller species. Research on North American salamanders has shown that interspecific competition is largely size-related in this way – larger larvae inhibit the growth, prolong the development and reduce the survival of smaller larvae (Stenhouse *et al.* 1983). Similar experiments to test for interspecific competition in European newts and salamanders are very few. Under laboratory conditions Crested Newt larvae inhibit the growth of Smooth and Palmate Newt larvae when food is limited. A similar effect was not detected in natural ponds, however, possibly because of a superabundance of food (Griffiths *et al.* 1994). Smooth and Palmate Newt larvae sometimes coexist in the same pond but are extremely difficult to distinguish in the field. The distribution of these two species is related to water quality. Smooth Newts are characteristic of hard-water ponds, while Palmate Newts seem to have a greater tolerance of soft water. However, in some parts of the country both species may coexist in the same pond, although rarely in equal numbers. This raises the question of whether water quality determines which species has the competitive edge. Experiments which have tested the growth of Smooth Newt larvae in soft-water ponds and Palmate Newt larvae in hard-water ponds refute this suggestion. The two species do equally well in terms of larval development under either set of conditions, suggesting that interspecific competition between these two very similar species is not influenced by water quality. It also suggests that the absence of one or other of the two species from a pond in a region where both naturally occur is unlikely to be due to competitive exclusion.

Larval newts and salamanders may well compete with other aquatic animals which feed on the same food types. These are likely to include predatory fish and carnivorous insects, such as water beetles, bugs and dragonfly nymphs. Although there have been no detailed studies of such interactions, these other predators may also include amphibian larvae in their diets. Both competition and predation may therefore play a role in interactions between them.

HABITAT SELECTION

One way different species of newts and salamanders coexist is by inhabiting different parts of the same pond. Szymura (1974) studied the spatial distribution of four species of newt larvae by netting different areas of ponds in Poland. All four species may live in the same water body (i.e. Smooth, Carpathian, Alpine and Crested Newts), but the degree of spatial separation between them depended largely on the range of habitat types that was available within a single pond. If the water body is a very uniform environment then there is little opportunity for newt larvae to differ in habitat selection.

Studies in Norway have shown that there is clear partitioning between Smooth Newts and Crested Newts within the water column (Dolmen 1983). In aquaria under semi-natural conditions, Smooth Newts are mainly bottom-dwellers whereas Crested Newts are found swimming within the water column. This vertical stratification is also reflected in their diets: Smooth Newts take more bottom-dwelling (or benthic) prey compared to Crested Newts, which prey mainly on mid-water (or nektonic) organisms (Dolmen & Koksvik 1983). Interestingly, this partitioning within the water column is the opposite of that observed in the adult stages, where Crested Newts are the bottom dwellers and Smooth Newts are more evenly distributed (Griffiths & Mylotte 1987).

There is little information available on habitat selection in stream-dwelling amphibian larvae. Compared to fishes, stream-adapted larvae are weak swimmers and habitat selection centres around avoiding being swept away by strong currents. Brook salamander larvae are therefore most often found in sheltered cracks, crevices or amongst stones. Fire Salamanders also sometimes breed in streams but their larvae seem to have little resistance to drifting. Stretches where the current is slack and there are hiding places under stones or amongst vegetation are therefore favoured (Zakrzewski 1970). Thiesmeier (1987) trapped drifting salamander larvae along a stream in Germany. The number of larvae captured was related to the flow rate of the stream, but large larvae were less prone to drifting than small larvae.

ACTIVITY PATTERNS

Seasonal activity of larvae is largely dictated by when the adult stages lay eggs or deposit young, but is also influenced by light and temperature. Daily activity rhythms are not clear-cut, although most species have larvae which are most active at night (Himstedt 1971). Towards metamorphosis daily activity rhythms seem to be lost completely, and a new rhythm appears some time after leaving the water. Fire Salamander and Crested Newt larvae avoid light more than Smooth Newt larvae, and this is consistent with the activity patterns of the adult stages.

Unlike frog and toad tadpoles newt and salamander larvae do not form social groupings. Any apparent aggregations within a water body are therefore more likely to reflect habitat differences rather than social behaviour.

NEOTENY AND PAEDOMORPHOSIS

In certain populations of newts and salamanders metamorphosis occurs only partially or not at all, and the animal continues to grow, and even breed, while retaining larval characteristics. The degree to which this occurs varies between species, as well as between populations of the same species, and a confusing terminology has consequently arisen. Strictly speaking, the term *neoteny* refers to the attainment of sexual maturity by juveniles. This is distinct from *paedomorphosis* which is the retention of juvenile features in the adult. In practical terms, the two are difficult to tell apart, and here the term *neoteny* is used in the broad sense suggested by Duellman & Trueb (1986), to encompass any situation where reproduction occurs during the larval stage. *Partial neoteny* is when the larva fails to metamorphose and continues growing, but does not actually breed in this state. *Total neoteny* occurs when the animal actually achieves sexual maturity and breeds in the larval state.

All families of the class Caudata have some neotenous members, but some species can be induced to metamorphose under certain environmental conditions. Four families, however, are exclusively neotenous (Cryptobranchidae, Sirenidae, Amphiumidae and Proteidae). Members of these groups always display total neoteny and can never be induced to metamorphose. The only European example of this condition is the Blind Cave Salamander (*Proteus anguinus*). There are two subspecies, one of which lacks pigment entirely, while the other is darkly pigmented and occasionally found in surface waters. If the pigment-less form is brought into bright light for several weeks, however, the skin may darken and the animal may eventually become almost totally black. As it lives in total darkness, the eyes have become reduced to light-sensitive spots covered by skin. Interestingly, the eyes of newly deposited young are more fully developed than those of adults, but a steady degeneration is observed over the first year or so.

Neoteny also occurs in several species of newts, and as far as the widespread species are concerned, the incidence of the phenomenon seems to decrease in the order Alpine Newt, Smooth Newt, Palmate Newt, Crested Newt (Fig. 5.4). In the Alpine Newt there is a tendency for entire populations to be neotenous. Some 22 neotenous populations have been reported from the Balkan States (Dzukic *et al.* 1990). The most well-known example is in Bukumirsko Lake in Montenegro, and has been assigned to a separate subspecies, *Triturus alpestris montenegrinus* (Radovanovic 1951). Other neotenous Alpine Newt populations have been reported in Italy and the south of France. In the Smooth Newt neotenous individuals usually turn up as exceptional individuals in otherwise normal populations. Out of some 1500 newts sampled from ponds near York, for example, only one was neotenous (Bell & Lawton 1975). Smooth Newts which display the condition are often partially

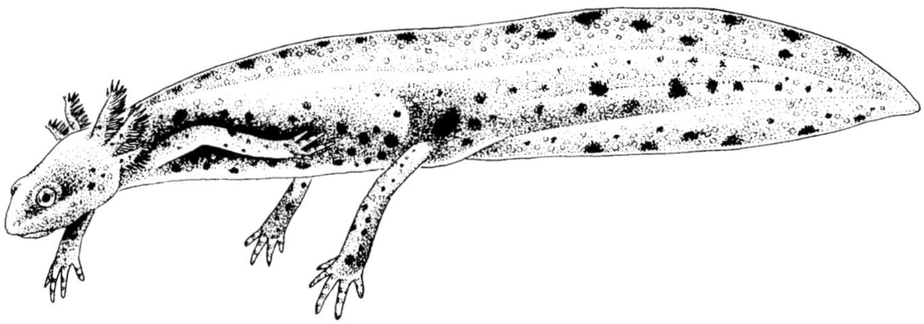

FIG 5.4 *Neotenous Alpine Newt. P. Benson.*

neotenous, and retain most of the larval characteristics while continuing to grow without metamorphosis. The condition is frequently associated with a lack of pigment, suggesting that the pituitary gland, which regulates pigmentation, is abnormal. Occasionally Smooth Newts are found which possess most of the adult characters yet retain the larval gills. These totally neotenous animals have been reported in both sexes and may breed successfully (Walhovd 1975, Benson 1979).

Neotenous Palmate Newts are known from Holland, Germany and France. Although neoteny in this species seems to be somewhat rarer than in the Smooth Newt, like the latter species both partial and total neoteny can occur. The most detailed study is that by van Gelder (1973) on a Palmate Newt population in Holland. This population comprised animals which were both partially and totally neotenous. The neotenous newts consisted of those larvae which had stayed in the water over two winters, and reached sexual maturity in their third year. As newts which had metamorphosed returned to breed at 3 years old, both normal and neotenous adults were found breeding in the pond at the same time.

Neotenous Crested Newt populations have been reported from only two areas; one in Scandinavia and the other in Montenegro. The Scandinavian population (*Triturus cristatus*) was located in Skrattabbor tarn in north-central Sweden, and comprised Crested Newts that were totally neotenous (Gislen & Kauri 1959, Dolmen 1980). This tarn is at the northern limit of the species' range, and also supported a population of neotenous Smooth Newts. The condition was almost certainly environmentally determined as the tarn is deep and situated at 500 m above sea level, in a region where the climate is cold and dry. Unfortunately the introduction of trout to the tarn in the 1960s is thought to have led to the extinction of this unique population. In Montenegro about 10% of the Alpine Crested Newts (*Triturus carnifex*) from Lokanj pond were totally neotenous. These animals were all males in breeding condition. A second water body nearby, Ceklin pond, contained no neotenous Crested Newts (Kalezic *et al.* 1994).

The regulation of metamorphosis in amphibians is under the control of the thyroid gland. This secretes a hormone, thyroxine, which stimulates metamorphosis. As neotenous newts can be induced to metamorphose by administering thyroid extract, the condition may be associated with a thyroid deficency. The neotenous Alpine Newts of Bukumirsko Lake have less active thyroid glands than normal newts, which suggests that any such deficiency may be environmentally determined (Pocrnjic 1979). Indeed, some Alpine Newt populations which are neotenous occur in deep water bodies at high altitudes. In the Balkan States 15 out of the 22 neotenous Alpine Newt populations occur in high altitude glacial lakes and tarns (Dzukic *et al.* 1990). Under such conditions larval development may be slowed considerably, and it may be beneficial to abandon the terrestrial stage altogether. The effect of thyroxine is also reduced in cold water. This effect, coupled with the fact that large, deep ponds may provide a relatively stable environment, may favour the evolution of neoteny. Alternatively, a dry terrestrial environment with a shortage of food or hiding places may favour an extended phase in water. The two Crested Newt ponds described by Kalezic *et al.* (1994) were located in dry, limestone country, and the newts at both sites spent more time in the water than those which occur further north in Europe. However, the pond which supported neotenous newts at Lokanj displayed less fluctuation in water levels than the pond at Ceklin which had only normal individuals. The newts at Lokanj also grew faster and had longer larval periods than those at Ceklin. The difference in reproductive strategy between the two populations is therefore consistent with the stability of the aquatic environments in which they live. The incidence of neoteny may therefore reflect the relative harshness and stability of the aquatic and terrestrial habitats. However, this cannot be the whole story as neoteny is not confined to harsh terrestrial habitats or even permanent ponds. In Italy, for example, neotenous Alpine Newts can be found in temporary ponds that eventually dry up, and in other areas normal newts are found in deep, permanent, upland ponds. Similarly, in many parts of Greece normal newts are found in ponds surrounded by dry, barren terrestrial habitat, features which would seem to favour neoteny. An alternative explanation is that neoteny allows newts to exploit a wider range of microhabitats within a water body than would be possible in normal forms. Whereas neotenous Alpine Newts in deep ponds have been observed to use the whole water column for foraging, metamorphosed Alpine Newts generally confine their hunting for food to the pond bottom. In this case neoteny may enable Alpine Newts to exploit a wider range of foraging habitats.

Metamorphosis

Following several weeks of steady growth, the final metamorphosis to a terrestrial newt or salamander can occur in as short a time as a week. The first signs of metamorphosis are a reduction of the tail fin and thickening of the limbs. Larval Fire Salamanders are cryptically coloured as larvae, but in the terrestrial phase adopt a striking black-and-yellow warning coloration. As the tail fin is absorbed, Fire Salamander larvae show the first signs of the change in colour pattern, although the colour pattern is not fully developed until some time after metamorphosis. Changes in the shape of the head accompany the transition from aquatic to terrestrial foraging. The tongue develops into an organ for catching food on land. The eyes change from a lateral to a dorso-lateral position, develop eyelids which can open and close, and become more prominent. The external gills are resorbed, and finally the young tailed amphibian crawls out onto land (Fig. 5.5). Newly metamorphosed newts and salamanders are sometimes referred to as *efts.* At this stage it is impossible to distinguish the sexes, and until adulthood is reached juveniles generally resemble females in colouration.

Very little is known about the eft stage in nature, partly because these tiny amphibians retire to inaccessible hiding places, sheltered from predators, temperature extremes and

FIG 5.5 *Growth and metamorphosis in the Fire Salamander. P. Benson.*

desiccation. Indeed, all of the small-bodied newts and the terrestrial salamanders will not return to water until it is time to breed, which can be several years later. One theory is that efts are excluded from ponds and lakes by the abundance of aquatic predators which would find them easy pickings. Young Crested Newts, however, only spend the winter months on land and migrate into ponds with the adult stages each year. The toxins present in the skin of Crested Newts probably provide protection against predators, and thereby enable the subadult stages to take advantage of the rich food resources present in ponds in the spring and summer (Bell 1979).

CHAPTER 6

Population and Community Ecology

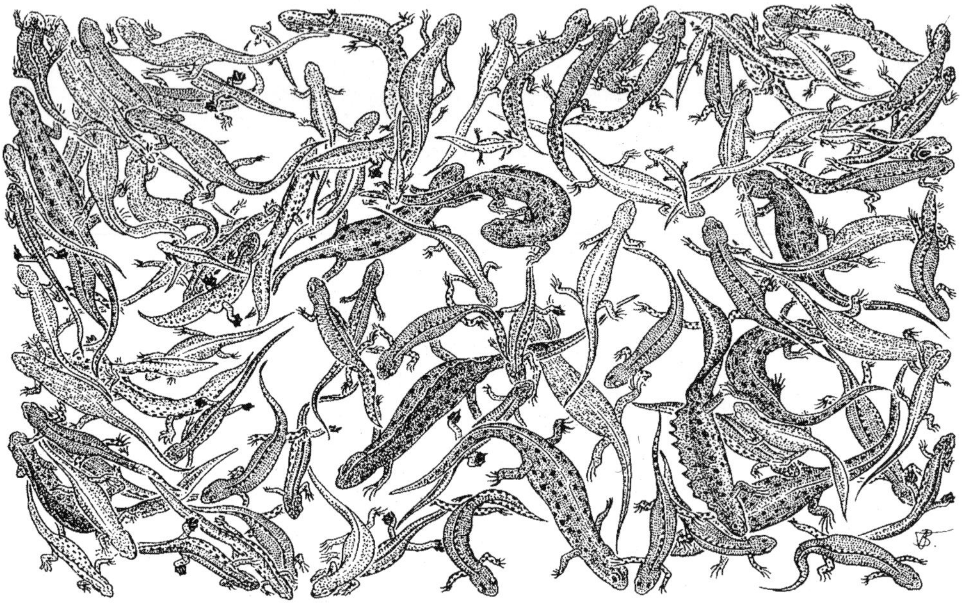

Triturus vulgaris/T. helveticus/T. alpestris/T. cristatus ♂♂ & ♀♀ and some larvae

WHAT factors regulate the abundance of newts and salamanders? This question may seem fundamental to understanding the conservation of these animals, but remarkably, for the vast majority of European species we have little basic information on population ecology. To understand what causes the year-to-year ups and downs in animal numbers we need information on two aspects of their ecology.

First, we need data on life history. This consists of information on how many animals are breeding, how many eggs they lay, how many larvae hatch, and how many animals survive at each stage of the life cycle. If we combine this with information on how old the animals are when they breed and how often they breed, we can construct a *life table.* This can then be used to predict the population size over a series of generations and tell us if the population is increasing, decreasing or stable.

Secondly, we need information on the factors which influence survival during each phase of the life cycle. These factors include predation, competition and disease. The effectiveness of these in reducing numbers will depend upon the density of the population. Competition for food, for example, will be much more acute at high densities of larvae than it will be at low densities. Similarly, a parasite or pathogen may be transmitted much more easily under crowded conditions. Another group of factors which will cause mortality at each stage of the life cycle are associated with the climate. Drying ponds, fires, flooded streams and cold winters can all have a devastating effect on the survival of newts and salamanders. However, the effectiveness of such climatic factors will not depend upon the density of the population. If a pond dries up, for example, there will be 100% mortality of the larvae, whether or not they were at high density or low density. Identifying the factors which lead to mortality is important because it can give a clue to the type of population trends that will occur, and vice versa. In general, density-dependent factors will maintain populations in a steady state. If the population exceeds that which can be supported by existing space or food then competition, predation or disease will become more intense and bring the numbers back down. Likewise, if the population is below its carrying capacity, the same factors will be relaxed, and allow the population to grow. On the other hand, density-independent factors are more unpredictable, and can result in catastrophic crashes in population size. Over a period of time the population trend will be one of peaks and troughs, corresponding to alternating periods of favourable and unfavourable conditions. In sum, competition, predation and disease will regulate populations so that they are relatively stable over time, whereas climatic factors (or any others that act in a catastrophic fashion) will result in a population with marked fluctuations.

Of course, in nature the situation is not quite this simple as both types of factors invariably have some bearing on population size. As a pond dries, for example, salamander larvae may become more crowded and competition for food more intense. If the pond is then refilled by rain those which survived the crowding in the drought will now be released from competition and will grow much better.

Complex Life Cycles

In passing through the stages of egg, larva, eft, immature and adult, a newt or salamander uses a series of dramatically different ecological niches during its life. This means that population regulation can occur at one or more stages of this *complex life cycle.* Predation or pond desiccation during the larval phase, for example, may place an upper limit on the number of animals that will later filter through to the adult population. If so, the adult stages may actually exist at a level well below the carrying capacity dictated by, say, their own food supply. Alternatively, competition for food between adults may reduce the number of eggs that each female can lay. This will mean that the larval population will ultimately be controlled by the food supply available to adults. A third possibility is that regulation will occur during two or more phases of the life cycle. Understanding what the regulatory factors are and where they act in the life cycle therefore provides further clues towards knowing what controls the abundance of newts and salamanders in nature.

Population Fluctuations

Very few studies have been conducted for long enough to determine how variable newt and salamander populations are over time. The longest running studies have been made in

North America. Mole Salamanders (*Ambystoma talpoideum*) were counted as they returned to breed each year at a pond in South Carolina, and showed considerable year-to-year fluctuations over a 12-year period (Pechmann *et al.* 1991). These fluctuations in the adult population were probably related to larval survival in previous years, which in turn was dictated by how long the pond lasted before drying. In the Appalachians, six species of salamanders have been studied for a period of over 15 years, and also display marked fluctuations (Hairston & Wiley 1993).

In Europe, one of the first studies to show a wide year-to-year fluctuation in the number of newts migrating to a pond was that by van Gelder (1973). This study monitored the numbers of Palmate Newts (*Triturus helveticus*) captured in a series of pitfall traps at a pond in Holland. Although there was a general upward trend in numbers over a 6-year period, numbers of newts captured varied from about three per trap in 1964 to over 30 per trap in 1969. More recently the numbers of Smooth and Palmate newts migrating to a pond in Wales were monitored from 1981 to 1985 (Griffiths *et al.* 1986). Palmate Newts were the most abundant species every year, but both species displayed two- to four-fold fluctuations in numbers over the 5-year period (Fig. 6.1). Interestingly, the population fluctuations displayed the same general trend in both species – 1982 was a very good year for both species, 1984 a particularly poor year. This suggests that whatever is controlling the numbers that migrate to the pond is affecting both species in the same way. Similar results were obtained in four species (Smooth, Palmate, Alpine and Crested Newts) inhabiting a series of ponds near Bonn in Germany (Blab & Blab 1981). Between 1976 and 1979 all four species displayed year-to-year fluctuations in population size but the general trend was similar in all the species. Another study on Smooth and Alpine Newts showed that larger populations seemed rather more stable than small populations (Glandt 1982). Although all these studies were of rather short duration, it seems likely that, as in the North American salamanders, fluctuations in population size may be related to climatic factors influencing the recruitment of individuals to the adult populations.

FLUCTUATIONS IN LARVAL SURVIVAL

If the recruitment of larval amphibians to the terrestrial stage varies from year to year according to climatic factors, fluctuations in the adult populations may be a reflection of larval survival. Various pieces of information on larval ecology support this theory. In some years entire cohorts of larval Smooth Newts may be destroyed by predation or pond desiccation, and such disastrous mortality may be commonplace in the life of this species (Bell & Lawton 1975). Crested Newts differ from the smaller newts in that juveniles return to water each year. The proportion of juveniles in a population may therefore give some idea of how successful metamorphosis has been in previous years. Again, the evidence is that larval survival is very variable. In Oxfordshire the percentage of aquatic juvenile Crested Newts varies from 0% in some ponds to as high as 71% in others (Bell 1979). At a pond which regularly dries up in mid-Wales, no juvenile Crested Newts were observed over a 3-year period. A series of dry years could therefore eventually lead to the decline of the adult population. Variation in the number of young newts in the population can also occur on a year-to-year basis within the same pond. At a permanent pond in Kent supporting about 100 adult Crested Newts, no larvae or juveniles were captured in 1991 or 1992 despite

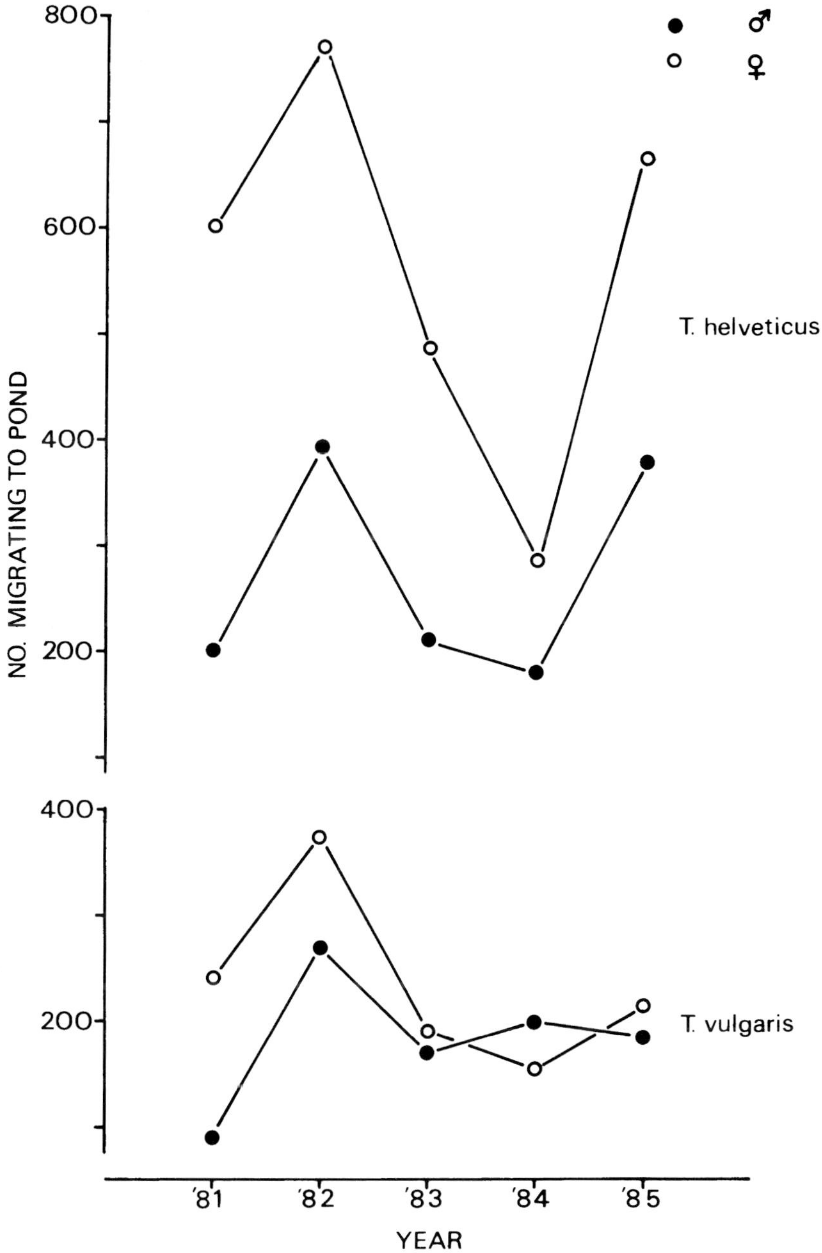

FIG 6.1 *Population fluctuations in Palmate (top) and Smooth (bottom) Newts at a pond in mid-Wales. Figures show the numbers of newts captured migrating to the pond over a 5-year period. Solid circles, males; open circles females. From Griffiths* et al. *(1986).*

extensive netting and trapping. In 1993, however, both larvae and recent metamorphs were regularly captured at the same site. The factors underlying these variations are unknown, but the fluctuations suggest that ups and downs may be a natural part of the newt's way of life.

If catastrophies form a natural part of their way of life, why do newt and salamander populations not rapidly become extinct? Laying extremely large clutches of eggs is one adaptation to cope with high larval mortality, but cannot explain how populations persist in the face of 100% losses in some years. Two further features of the life-style help deal with this problem – longevity and colonisation.

LONGEVITY

The longer a newt or salamander lives, the more chances it will have to breed. If animals breed repeatedly over a number of years, it will not matter too much if one, or even a few, larval cohorts are completely lost to unpredictable catastrophies. Failure in one year may be offset by success in the following year. By breeding more than once, newts and salamanders thereby increase the chances of at least some of their offspring surviving.

In recent years a new technique, known as *skeletochronology*, has enabled biologists to determine the ages of amphibians very accurately. Just as trees lay down growth rings each year, the limb bones of amphibians have a resting phase in their growth during the winter months. When the bones are sectioned and examined under the microscope, these resting phases can be counted as the number of darkly staining rings. The number of rings will therefore indicate how many winters the animal has experienced and consequently how old it is (Fig. 6.2).

Compared to other animals of similar body size, many newts and salamanders are long-lived. In Britain Smooth and Palmate newts return to breed when they are 2–3 years old and may live for up to 6 years (e.g. Verrell & Francillon 1986). The larger bodied newts are even longer lived. Crested Newts can breed for the first time at 2–3 years old, Marbled Newts a year or so later. Both species can live for 14 years or longer (Francillon-Vieillot *et al.* 1990). A female newt will therefore breed several times during her lifetime. Fire Salamanders have a similar life span. On Mt Carmel in Israel females return to breed for the first time when they are about 6 years old, and will breed in every subsequent year until they are 10–11 years old. Some salamanders may live for as long as 20 years (Warburg 1992, 1994).

In colder areas growth and development is slower, and newts seem to reach sexual maturity later but live for even longer. In Sweden Smooth Newts breed for the first time 2–3 years later than they do in in Britain, and can live for up to twice as long (Hagström 1979). Alpine Newts collected from a pond in Bohemia at an altitude of 490 m breed for the first time at 3 years old, with the oldest individuals being aged 10 (Smirina & Rocek 1976). In the Austrian Alps at an altitude of 1643 m, however, no Alpine Newts breed before the age of 10, and some individuals live for longer than 20 years (Schabetsberger 1993).

Interestingly, in most species studied to date males reach sexual maturity a year or two before females. In terms of reproductive success, it is in the interests of males to start breeding as soon as possible. As far as the female is concerned, however, the number of eggs she can carry is limited by her own body size. By delaying sexual maturity by a year

FIG 6.2 *Section through the humerus of a 12-year-old female Marbled Newt, showing growth rings. White bars indicate the lines of arrested growth formed each winter (J. Castanet).*

or so she can grow a little more and therefore carry more eggs when she does return to breed.

Most newts and salamanders therefore breed every year for several years. This means that if a disaster should completely eliminate all eggs or larvae in a pond, the adults will have further chances to breed in another year. The longer lived the adults are, the more resilient they will be to catastrophies.

COLONIZATION

Of course, if catastrophies occur on a regular basis the population will decline and even if the adults are long-lived, the population could become extinct. A series of dry years, for example, could lead to the eventual elimination of a newt or salamander population. If rainfall patterns then return to normal there may well exist suitable ponds with unfilled niches for newts and salamanders. This sequence of events may well explain why there often appear to be ideal breeding sites for amphibians in some areas which are in fact not used by the appropriate species. The only way rejuvenated or newly created sites can start supporting amphibian populations once again is if they are recolonised.

Despite apparently being poorly adapted for life on land, newts and salamanders, and indeed amphibians in general, are remarkably adept at colonising new ponds. Again, some of the evidence for such abilities is based on circumstantial or anecdotal information, but the evidence is compelling. A garden pond in mid-Wales was colonised by all five native amphibian species, including Smooth, Palmate and Crested Newts, within 18 months of its construction. This pond adjoined an area of woodland and pasture, which had several amphibian ponds, the nearest of which was over 200 m away. Smooth, Palmate and Crested Newts in mid-Wales have also managed to colonise some fairly remote upland pools, a kilometre or more from the nearest breeding populations. A much more quantitative study of colonization was carried out over a total period of 15 years on a newly created pond in northern France (Arntzen & Teunis 1993). The site was colonised by newts the year following its construction, probably by about four individuals. These animals are thought to have migrated from three neighbouring ponds between 120 m and 360 m from the new pond (see case study below).

Natural colonization may therefore be one way newts and salamanders can start up new populations in new ponds or ponds where extinction has occurred as a result of catastrophies. However, successful colonization does rely on (1) suitable donor populations being present nearby, and (2) the terrestrial habitat between the sites being suitable for the animals to migrate across. It is therefore unlikely that a pond will be colonised naturally if the nearest other pond is some distance away, or if it is cut off from a donor site by unsuitable habitat.

It is perhaps inappropriate to regard the newts and salamanders which exist at an individual breeding site as self-sustaining populations. Because of the unpredictability associated with water bodies the amphibian life-style relies on some interchange of individuals between neighbouring ponds, to offset any chance population crashes which may occur at any one site. It is much better then to regard a viable population as one which has a number of subunits focused on individual breeding areas, between which exchange of individuals can occur. This type of population is referred to as a *metapopulation.* As discussed in Chapter 8, understanding how habitat fragmentation can disrupt the structure of newt and salamander populations is crucial to developing ways in which to conserve them.

SEX RATIO

The number of eggs or young produced each year is also influenced by the relative numbers of males and females in the population which are capable of breeding. In the vast majority of vertebrates the sex of the offspring is genetically fixed early on during development, and this usually ensures that the same number of males and females will be hatched or born into the population. The Spanish Ribbed Salamander (*Pleurodeles waltl*) is an exception in that the sexes can be reversed by raising the temperature from 20°C to 30–32°C during larval development (Dorazi *et al.* 1995). However, there are many other ecological factors that can alter the ratio of males to females in a population of newts or salamanders. First of all, there may be differences in the survival between males and females. In the mountain population of Alpine Newts studied by Schabetsberger (1993), for example, the sex ratio averaged one male for every four females. Males seemed to suffer much higher mortality when migrating to the lake, and had an annual survival of about 40% compared to over 70% in females. Studies which have monitored breeding migrations of newts by surrounding a pond with a drift fence and pitfall trap system invariably produce a female-biased sex ratio in immigrating newts in the range of one male for every two to three females (Harrison *et al.* 1983, Verrell & Halliday 1985, Griffiths *et al.* 1986). However, far fewer animals are captured leaving the pond than entering, and interestingly, the proportion of females is usually lower. One explanation for this is that females suffer higher mortality in the pond than males. Although different methods clearly can yield different results, it does seem that the sex ratio is variable in newt populations, although in general there are not extreme deviations from 1:1.

Differences in the age structure of a population will also influence the sex ratio. As male newts seem to reach sexual maturity a year or so before females, this may mean that there are more males than females in the breeding population at any one time. Even if there are no differences between the sexes in terms of survival or age structure, there may be apparent differences in sex ratio which are related to their behaviour. By staying in the water for as long as possible, a male stands a better chance of mating with a large number of females. A female, however, has a fixed number of eggs or offspring she can produce each year, and having deposited these there may be no advantage in staying in the water. If males therefore stay for longer in a pond than females this can give the impression that there are more males than females. In a Smooth Newt population in London, the sex ratio of terrestrial males:females was about 1:3 for most of the year. In water, however, the ratio was closer to 1:1. This could be explained by the fact that males spent longer in the breeding ponds than females (Griffiths 1984).

In the more terrestrial Fire and Alpine Salamanders, males are often more abundant after rain. Again, this is probably related to the fact that males are actively searching for females at this time of year and are more conspicuous, coupled with an earlier maturation.

Coexistence of Species

The habitat where a species lives, the food it eats, and the times when it is active define its ecological niche. When several similar species live in the same area, the traditional view has been that they must use different ecological niches within that area in order to coexist. Species with very similar niches may compete for space or food, which will result in one or other of them declining in abundance, or even going extinct. In sum, an assemblage of different species of newts and salamanders must share out the available resources between them if they are going to coexist in the same area. Describing how coexisting species differ in their use of habitats, food or time has long been used as justification for how they manage to live together. Recent research has cast some doubt on the idea that 'resource partitioning' must occur if species are to coexist, and studies of newts and salamanders have been instrumental in showing that this is not always the case.

Within the genus *Triturus* there is considerable overlap between species in their geographical ranges. Indeed, a minimum of two newt species in any one area is the normal situation over most of Europe, with these species often occurring in the same pond. The highest species density occurs in northern and central France, where up to five species may be found (Arntzen & de Wijer 1989; Fig. 6.3). Although assemblages of two or three species in a pond is a relatively common situation, there are often size differences between them. Indeed, newt assemblages usually contain at least one example from each of the large-bodied and small-bodied groups. Examples are Bosca's and Marbled Newts in Spain; Palmate Newts, Alpine Newts (which are intermediate in size) and Crested Newts in France and Germany; and Italian Newts, Alpine Newts and Alpine Crested Newts in Italy. In general, the overlap in range of two species of similar size is narrow, although the distribution of Smooth and Palmate Newts in north-west Europe is a notable exception to this rule (Joly & Giacoma 1992).

Size differences between species usually reflect ecological differences (e.g. different diets) so may allow the coexistence of species which are otherwise closely related. A more difficult question is how newt species of similar size coexist in the relatively narrow areas where they overlap. In such cases, are there subtle differences in species ecology?

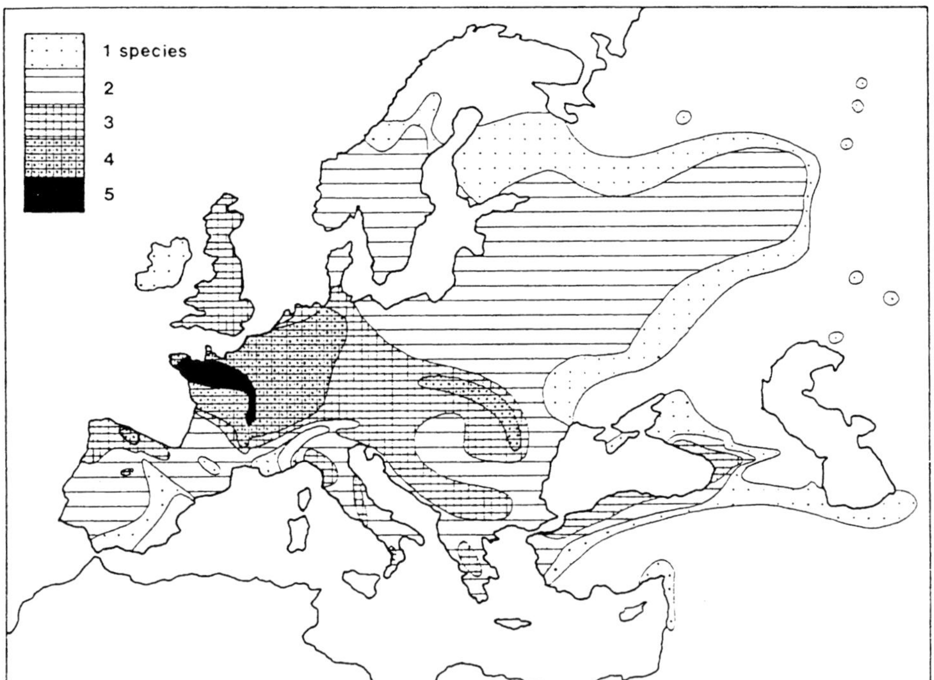

FIG 6.3 *The density of newt species found in different parts of Europe. In parts of north-west and central France up to five newt species may coexist in the same area. From Arntzen & de Wijer (1989).*

One aspect of habitat type that separates newt species to some extent is altitude. In Germany, the abundance of Crested and Smooth Newts decreases with altitude. On the other hand, Alpine Newts and Palmate Newts become more abundant with increasing altitude (Glandt 1980). Similar observations have been made in the Beskid Mountains of Poland, where the Carpathian Newt occurs. Here four species occur along a lowland--highland gradient in the order Crested, Smooth, Carpathian and Alpine Newt (Swierad 1980). What is clear, however, is that despite these altitudinal preferences, there are broad areas of overlap between the species. In the Beskid Mountains, for example, all four species may coexist in ponds between 400 and 600 m above sea level (Fig. 6.4).

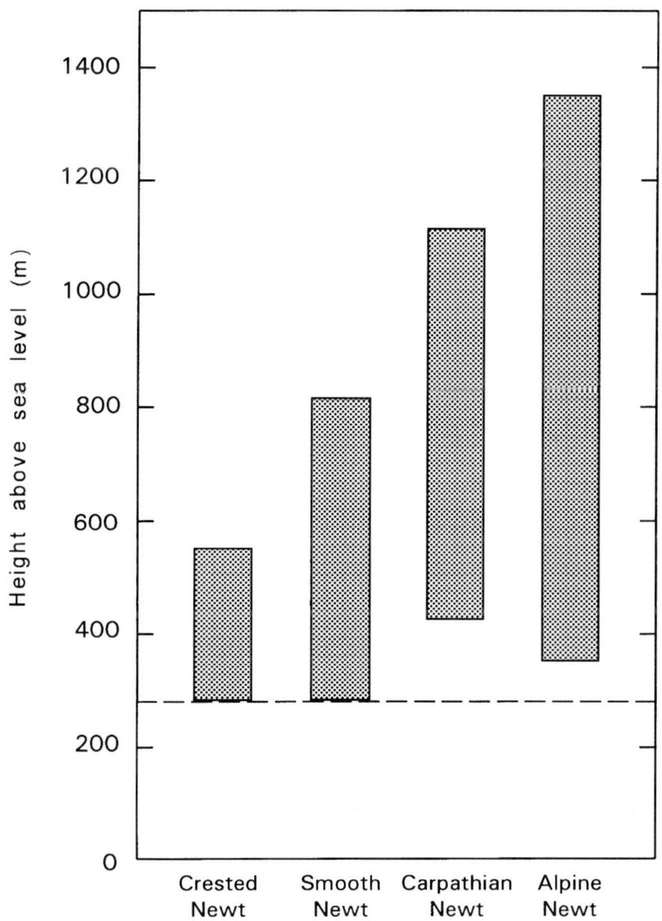

FIG 6.4 *The vertical distribution of four species of newts at altitudes of above 280 m in the West Beskid Mountains, Poland. Broken line indicates 280 m (lowest altitude surveyed). From Swierad (1980).*

Geology and water quality are other factors that influence the distribution of newt species. Smooth and Crested Newts are characteristic of hard-water areas that are rich in metals, particularly calcium. Indeed, the presence of Crested Newts is usually a good indicator that the pond will also contain Smooth Newts. On the other hand, the Palmate Newt is often associated with soft-water areas which are metal-deficient (Cooke & Frazer 1976). Once again, however, the separation between the species is incomplete. There are many areas where these three species overlap and even occur in the same pond.

If partitioning between the species by altitude or habitat type is incomplete, how do different species coexist when they use the same breeding pond? This could be achieved by breeding at different times, using different parts of the pond, or feeding on different prey.

Although there may be slight differences between newt species in the timing of their breeding activities, the time needed for larval development coupled with the temporary nature of many ponds means there is little scope for finely dividing up the short breeding period between species. Consequently adults of different species are likely to be breeding in a pond at the same time, and the resulting larvae will also overlap in their periods of development. When this occurs there may be slight differences between the species in terms of which parts of the pond are used. Adult Crested Newts, for example, are predominantly bottom-dwellers, compared to Smooth Newts which occupy a broader range within the water column. (Interestingly, the larvae of these two species are also partitioned to some extent but in a way that is the reverse of the adults – see Chapter 5.) These differences may serve to minimise interspecific predation or competition, but again the differences are not clear-cut: both species can often be found in the same part of the pond. Moreover, when Smooth Newts coexist with similar-sized Palmate Newts, there is absolutely no difference between the species in where they live within the pond (Griffiths 1987).

As discussed in Chapter 2, newts are generally unselective feeders, whose diet is limited mainly by prey size. Where two or more species coexist in the same pond, competition for food could be reduced by different species feeding on different prey types. Certainly, large-bodied newts will consume a wider variety of prey sizes than small-bodied newts, but again there seems to be broad overlap between these groups, and even higher dietary overlap within the size groupings. Indeed, at a pond in mid-Wales the diet of Smooth Newts was almost identical to that of Palmate Newts (Griffiths 1986).

Within newt assemblages then, different species may use the available resources in different ways, and this may facilitate coexistence between them. Large-bodied and small-bodied newts in particular seem to have different niches, which may serve to minimise the degree of predation or competition between them. Where two or more species of similar size live together, niche differentiation is much less clear. How can such a high degree of niche overlap be explained? The idea that an assemblage of different species must partition the resources between them hinges on an important assumption – that the resources are in short supply. If there is enough food or habitat to go round, there will be no need for different species to use different parts of the resource spectrum. Certainly, some ponds support an abundance of food which is unlikely to ever be depleted by newts to a level whereby it is in short supply. This probably explains how Smooth and Palmate Newts in mid-Wales can live in the same parts of the pond and eat exactly the same food (Griffiths 1986, 1987). So if food is superabundant, why do the newt populations not expand in number until food is in short supply? There are probably two reasons why this does not occur. Firstly, their complex life cycle means that the numbers of larvae which survive to metamorphosis put an upper limit on the number which can later return to breed as members of the adult population. If larval mortality is high, then it is quite possible that the number of adults will never be large enough to fully use all the food or habitat types that are available to them in the pond. Secondly, density-independent factors may also have a role to play. If severe winters, dry summers or other catastrophies are a common source of newt mortality, then it is unlikely that numbers will ever build up to a level whereby the available resources are saturated. Of course, larval mortality and the incidence of catastrophies varies enormously from year to year. Moreover, some regions will support pond communities which are much more stable than others. It seems likely, however, that the unpredictability of the newt life-style may be an indirect reason why several similar species can often co-exist in the same pond.

HYBRIDISATION

Considering that in many parts of Europe ponds contain at least two species of newts, there is the potential for interbreeding between different species. By definition, a species is one that will only produce fertile offspring if it mates with its own kind. In the majority of animals hybrids are therefore infertile and a waste of reproductive effort, which means that mechanisms for avoiding hybridisation are likely to evolve. Differences between species in reproductive behaviour usually mean that males or females respond only to the opposite sex if it is a member of the same species. In this way, 'mistakes' are avoided. Within the newts, however, there are a few exceptions.

In general, where the ranges of two or more newt species widely overlap, hybridisation is absent or very rare. On the other hand, where there are narrow areas of overlap hybridisation in newts may be more frequent. Within the Crested Newt species group, for example, hybrids may be formed where the ranges of two or more species overlap. As these species are quite closely related (and until recently were considered to be just subspecies of *Triturus cristatus*), such hybrids would be quite difficult to identify in the field. More intriguing are the hybrids that are formed between more distantly related species. One of the best described cases of this is the hybridisation that occurs between Crested Newts and Marbled Newts in north-west France. Crested Newts are a north-west European species, which reach their southern boundary in central France. The Marbled Newt on the other hand is a south-west European species, which reaches its northern limits in the same area. The two species therefore have a narrow band of overlap in northern and central France. The *départment* of Mayenne lies within this band, and here the two species have quite different habitat preferences. Crested Newts are found mainly in flat, open, low-lying areas, while Marbled Newts occur in hilly, wooded habitats. Certain areas have features which favour both species, however, and these may contain ponds which support both Crested and Marbled Newts (Schoorl & Zuiderwijk 1981). The proportion of Crested to Marbled Newts in such ponds varies, but hybrids between the two species can comprise up to 19% of the total number of adults (Arntzen & Hedlund 1990). Hybrids may mate with each other, or back-cross with either of the parent species, but the eggs they produce have low viability. Even within this narrow area of overlap, then, isolating mechanisms have evolved to ensure that the two species can maintain themselves and that the impact of the hybrids on the newt populations is minimised.

Another case of hybridisation occurs where the ranges of the Smooth Newt and the

Carpathian Newt overlap. Narrow areas of overlap between species have probably come about comparatively recently, and the low viability of hybrids in such areas means that interspecific matings will eventually be eliminated during the course of evolution. Where species have overlapped for much longer it is likely that evolution has already taken its course, and resulted in mechanisms which prevent matings between different species. This is probably the case in Smooth and Palmate Newts which have a large area of overlap in north-west Europe. The courtship behaviour of these two species is very similar, but tail fanning is faster and used more often in Palmate Newts (Halliday 1977). This behavioural difference, coupled with the different breeding characters of the males, seems a very effective way of preventing interspecific matings. Even so, newts sometimes still make mistakes. A newt captured in a Welsh pond in 1986 was found to be a Smooth × Palmate hybrid (Fig. 6.5). Examination of the chromosomes of this animal showed that it had resulted from a mating between a male Palmate and a female Smooth (Griffiths *et al.* 1987). A second hybrid has recently been discovered from a pond in France (J.W. Arntzen, pers. comm.). These two specimens remain the only definite hybrids between Smooth and Palmate Newts that have been discovered in nature.

Population Regulation

In general the population ecology of newts has been much better studied than that of the more cryptic salamanders simply because it is easier to observe animals which converge on breeding areas each year. Although no complete life tables have so far been produced for any species, some populations of newts and salamanders have been sufficiently well studied as to suggest those factors which may be important in regulating population size. Four such case studies follow.

Alpine Newts in Austria

Robert Schabetsberger's study of the population dynamics of Alpine Newts in Lake Dreibrüdersee in the Austrian Alps was combined with an investigation of the impact of newts on its principal prey, the copepod *Arctodiaptomus alpinus* (Schabetsberger 1993, Schabetsberger & Jersabek 1995). Lake Dreibrüdersee represents a very harsh environment, as the winter lasts some 7 months and the water becomes devoid of oxygen under a 3 m layer of snow and ice (Fig. 6.6). Adults start to migrate to the lake as the ice melts in June. The adult population ranged between 666 and 864 individuals during the breeding period in July and August. When food consumption of adult newts was compared to the abundance of copepods, it was found that the impact of newts on their principal prey was low. Larvae usually hatched in August, and over a 4-year period, the population size of late larval stages varied between 4000 and 26000. However, very few of these reached metamorphosis as most larvae died each winter as a result of deoxygenation of the water under the thick layer of ice. Larval survival is therefore highly variable, and only in 1 in 5 years is larval abundance likely to be so large as to have a major impact on the copepod food source. As adult Alpine Newts live for 20 years or more in this population, they will have several opportunities to breed, and recruitment to the adult population is only needed once every few years to maintain the population.

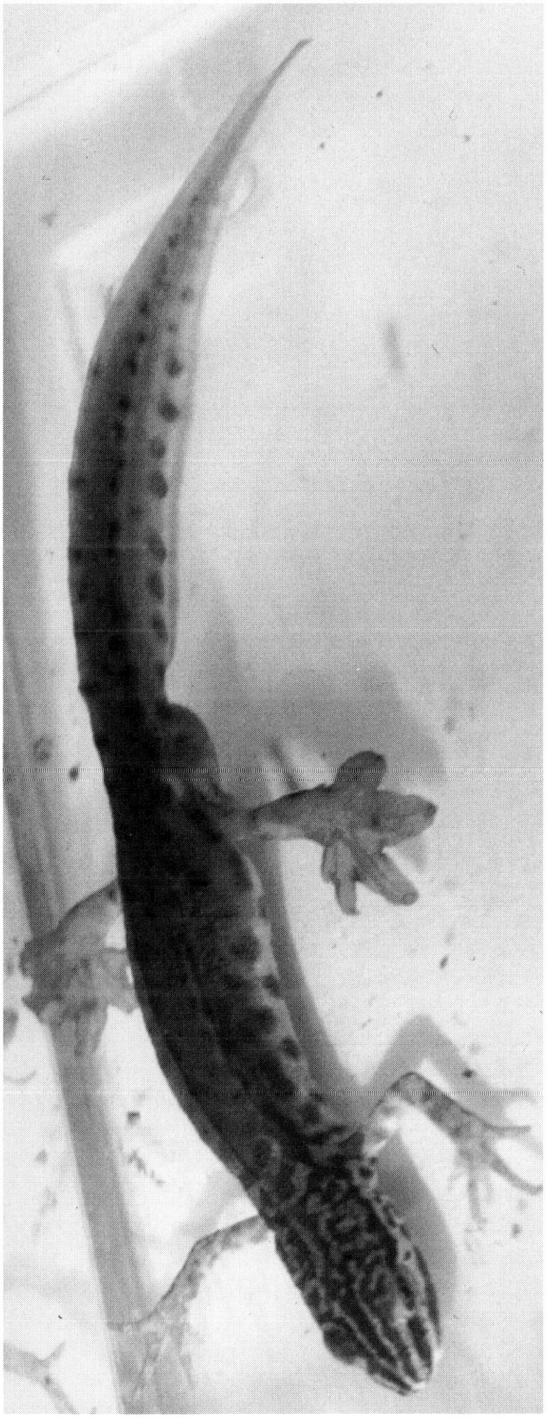

FIG 6.5 *Hybrid Smooth × Palmate Newt captured in a pond in mid-Wales. From Griffiths* et al. *(1987)*

FIG 6.6 *Lake Dreibrüdersee, an alpine tarn in Austria where ice cover for most of the year regulates the breeding population of Alpine Newts (R. Schabetsberger).*

FIG 6.7 *The distinctive belly patterns of Crested Newts allow different individuals to be identified in population studies. R. Griffiths.*

The most important factor limiting the population size of newts in Lake Dreibrüdersee therefore seems to be high larval mortality in winter. Recruitment to the adult population is consequently low, which means that adult newts are not limited by their food supply. Because of high larval mortality under the ice, only occasionally will food become limiting for larval stages.

Crested Newts in France

Pim Arntzen and Bas Teunis studied the development of a Crested Newt population in the Pas-de-Calais area of north-west France over a total of 15 years (Arntzen & Teunis 1993). Individual newts were recorded by photographing the patterns of black spots on the belly which are distinctive in different individuals (Fig. 6.7). The newly created pond was colonised by about four newts from a neighbouring pond in 1977. The population increased rapidly to 335 individuals in year 5 and then crashed again to about 16 newts in year 7 (Fig. 6.8).

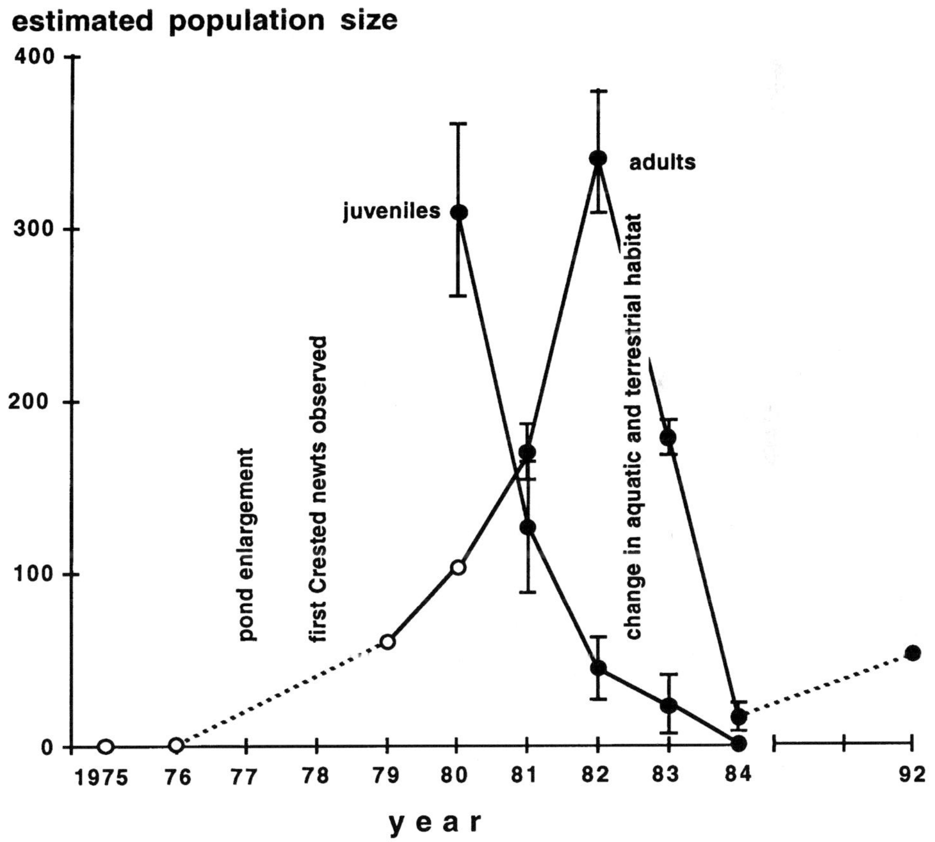

FIG 6.8 *Population fluctuations in Crested Newts in a pond in northern France. Solid circles, population estimates by mark–recapture methods; open circles, population estimates by other methods. Vertical bars indicate the standard errors of the estimates. From Arntzen & Teunis (1993).*

The population crash may have come about as a result of extensive alteration to the pond and its environs in 1983–1984. If these years of unusually high mortality are excluded, the annual survival of adults varied little with age and was about 49%. (This is somewhat lower than previously published estimates which put Crested Newt survival at between 50% and 78% per year.) The survival of juveniles varied between years, but averaged around 22%. Juveniles took 2–3 years to reach sexual maturity, and years in which the juvenile population was high were followed by years with large adult populations. The size of the juvenile population was in turn controlled by larval survival, which varied considerably between years. The reasons underlying the variation in larval survival are unknown, but predation, particularly by sticklebacks which were introduced to the pond in 1981, may be important. The population trends also suggest that those years when recruitment is low correspond to years with high adult population densities. This suggests that cannibalism of adults on larvae could also be occurring in a self-regulating way.

Smooth Newts in England

Graham Bell and John Lawton studied the population ecology of Smooth Newts in a series of ponds in Oxfordshire in the early 1970s (Bell & Lawton 1975). The survival of eggs to hatching was determined by two different methods. The first method involved taking cores of vegetation at regular intervals from a pond and searching it for newt eggs. By counting, ageing and staging all the eggs found it was possible to calculate an estimate of how many eggs survived to hatching. The second method was a more direct one, in which female newts were first allowed to lay eggs on fronds of vegetation in the laboratory. These fronds were then marked, placed in a natural pond, and the progress of the eggs monitored at regular intervals. Both methods gave similar estimates of survival to hatching of around 2.5%. However, the authors emphasise the year-to-year variability that may occur, and suggest that their estimates of egg survival may be lower than that which normally occurs. By taking samples of larvae throughout the season and analysing the number captured and their sizes it was also possible to arrive at an estimate of 10% survival from hatching to metamorphosis. Using more direct methods Verrell (1985) calculated survival from egg to eft as about 2% at another pond in southern England, and suggested that Bell & Lawton may have overestimated larval survival at their sites. Bell & Lawton also showed that most larvae metamorphosed in the same year as they hatched but a small proportion overwintered and metamorphosed the following spring. The annual survival of immature newts was 80%, and that of adults 45% in males and 55% in females (Bell 1977). Again, these figures may be inaccurate, as Bell made a number of assumptions in his analyses which may not be applicable. In particular he estimated the ages of newts in his populations from their size, a method which is now known to be unreliable. Certainly, his estimate that newts breed for the first time at an age between 3 and 7 years is somewhat later than more recent studies which have used the more reliable method of skeletochronology.

Population sizes were calculated by mark–recapture methods, and varied between 11 and 1298 individuals with a mean size of 72.5 newts. Despite the shortcomings of Bell's work, his view that the survival of eggs and larvae is generally low or erratic is supported by more recent data. Such unpredictable recruitment will lead to variation in the number of adults observed in different ponds in the same year, or the same pond in different years.

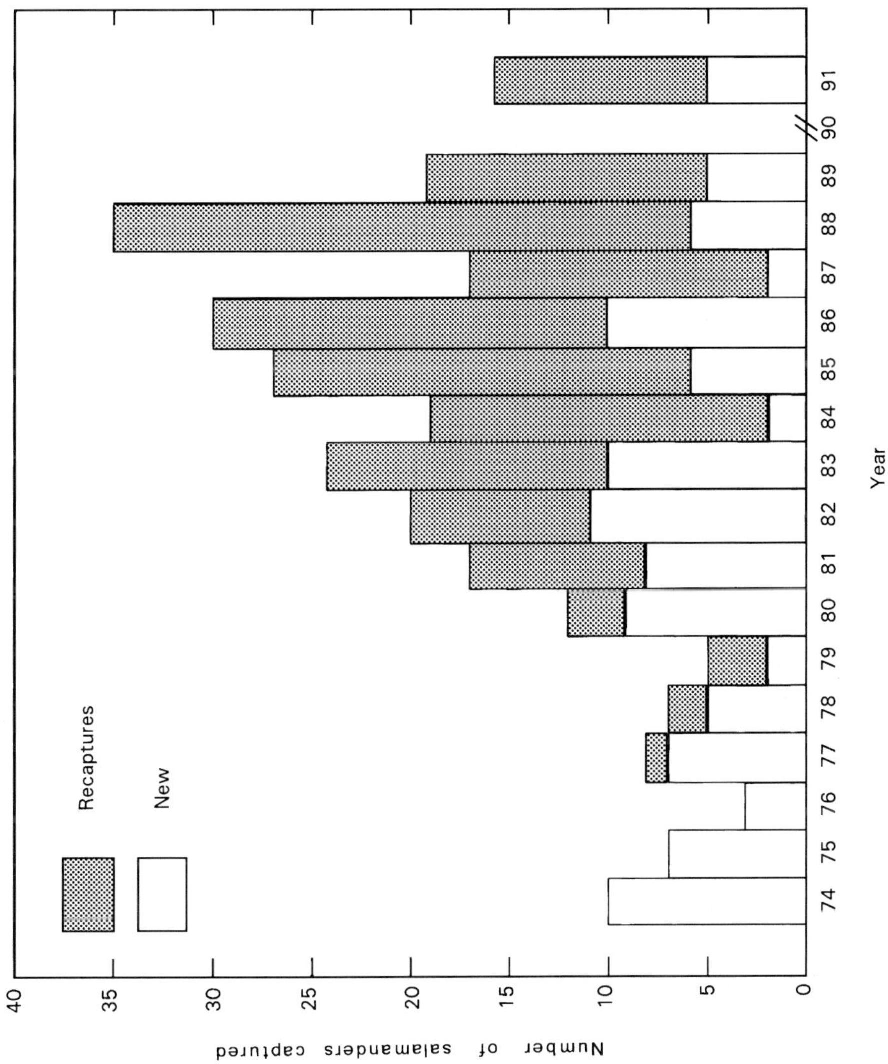

FIG 6.9 *The number of Fire Salamanders captured over an 18-year period on Mt Carmel, Israel. From Warburg (1994).*

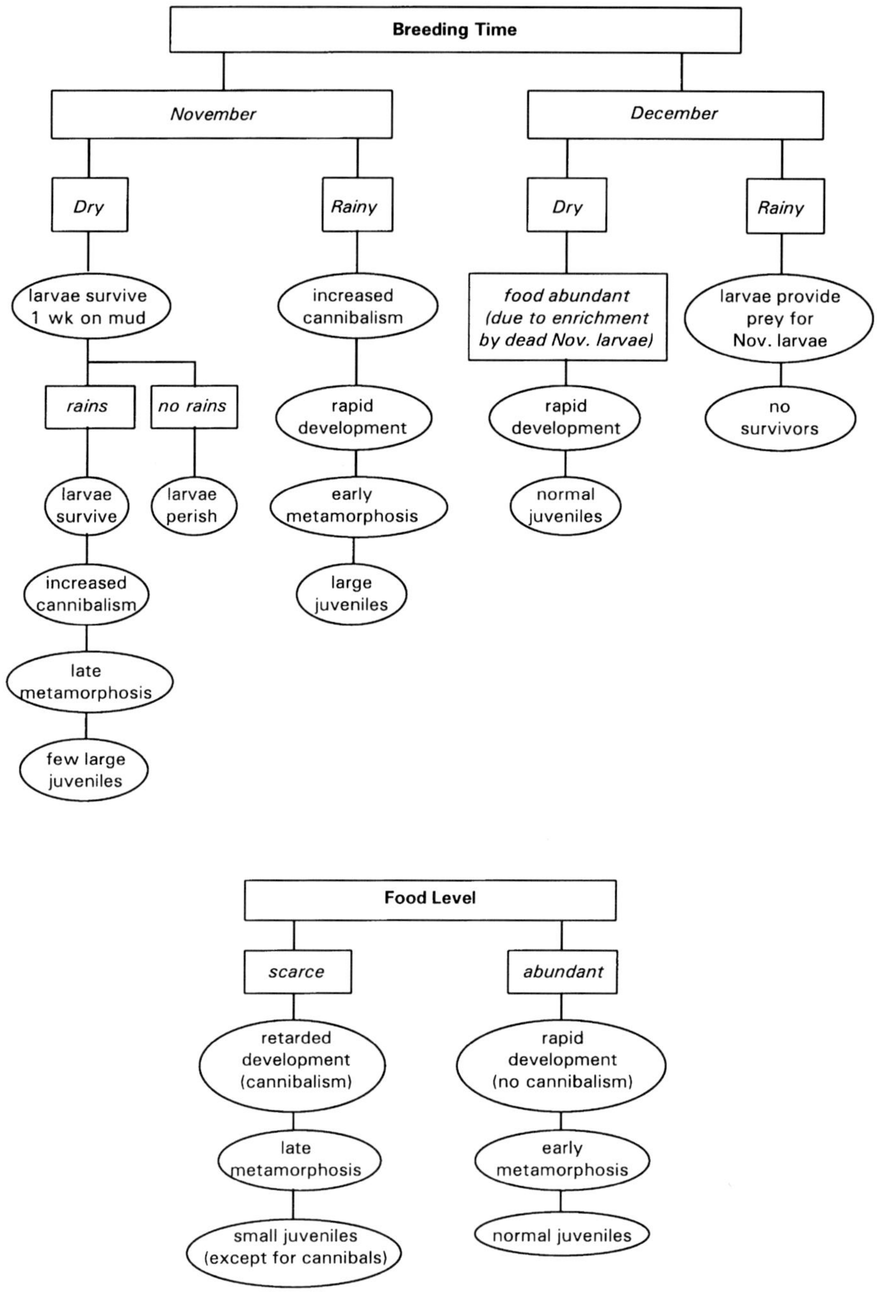

FIG 6.10 *Flow-chart showing the factors affecting the survival of Fire Salamander larvae in rock-pools on Mt Carmel, Israel. Boxes indicate the prevailing environmental conditions; ellipses indicate the effects on salamanders. Redrawn from Warburg (1992).*

Fire Salamanders in Israel

The Fire Salamander reaches the southern limits of its distribution area in Israel, where it is known from just three isolated populations. On Mt Carmel it has been studied by Michael Warburg and his colleagues for over 15 years. Here the climate is quite arid and unlike the more typical salamander habitats of north-west Europe. The only reason that Fire Salamanders can survive in this apparently inhospitable area is because the rainfall on the northern part of the mountain range is higher than average (Warburg 1986a,b). The salamander population here is rather small with only 131 different individuals being captured over a 18-year period (Warburg 1992, 1994; Fig. 6.9). This amounts to a density of only one salamander every 200–300 square metres. However, adult salamanders are relatively long-lived, with females reaching maturity at 6 years and then breeding for an average of four successive years. The success of the population in this suboptimal habitat seems to lie in its breeding strategy which is adapted to the pattern of rainfall. The earliest larvae are deposited in November but these usually die because the ponds dry up. In most years, then, larvae which are deposited after the ponds have refilled in December will have the best chance of survival. In a wet year, however, members of the November cohort will survive and by virtue of their larger size will cannibalise those larvae deposited later in December. On average this happens every third year. Wet years therefore favour early-breeding females which produce the cannibalistic early cohorts, while dry years favour the late-breeding females which produce the later cohorts. The early larvae which have cannibalised their smaller counterparts will give rise to larger sized salamanders at metamorphosis than those from later cohorts. Although early breeding will, on average, only give rise to successful metamorphosis once every 3 or 4 years, the larger juveniles that are produced have a better chance of survival on land. Breeding late means that larvae have a higher chance of making it through to metamorphosis, but the cost of this strategy is a smaller size at metamorphosis. The two strategies therefore have advantages and disadvantages which are related to the variable rainfall patterns, and are consequently both maintained within the population (Fig. 6.10).

There are some interesting comparisons to be made between this relic population and those in Europe. The Mt Carmel Fire Salamanders are larger in size than those elsewhere, with the largest females reaching nearly 30 cm in length. In relation to body volume, a large body size presents a smaller surface area across which body water can be lost, and this may be an adaptation to the dry conditions of Israel. Moreover, the number of larvae each female produces is much higher on Mt Carmel than in Europe. These large clutches may provide a further insurance against cannibalism and pond desiccation.

Another Fire Salamander population in Israel has been studied at Tel Dan, an area which is much wetter near the headwaters of the Jordan River. Compared to Mt Carmel, in this less extreme environment the density of salamanders is higher, the sizes of the individuals are smaller, and the number of larvae produced by each female is lower. Interestingly, these population characteristics more closely resemble those from typical European habitats (Degani & Mendelssohn 1982).

CHAPTER 7

Conservation

Urban development

WHY is it important to conserve newts and salamanders? As described in previous chapters, many species are strikingly attractive animals with curious, often bizarre life-styles, that we are only just beginning to understand. Moreover, the deep associations that these animals have had with folklore underlines the sense of wonder that newts and salamanders have instilled throughout human history. Are these good enough reasons to be concerned about them? On the down side, newts and salamanders are of little economic importance, which makes it difficult to express their value to humankind in monetary terms. As conservationists often point out, however, extraordinary sums of money are paid for the creations of the great artists, such as Rembrandt or Picasso, yet we knowingly put at risk creations which have been moulded by millions of years of evolution. If wildlife could be viewed in the same aesthetic light as works of art the job of conservation would be made a lot easier.

Certainly, there are sound ecological reasons for conserving newts and salamanders. These amphibians play a key role in the ecology of ponds and streams. As carnivores during both adult and larval stages, they influence the diversity and abundance of their invertebrate prey. This in turn has an effect on the plant communities upon which many of the invertebrates feed. On land, the impact of newts and salamanders on their invertebrate prey is less well understood. However, they do consume many invertebrates which are potential pests of plants of economic importance. Whether newts and salamanders can be used to control such pests is therefore worthy of future research. The complexities of these interactions are still to be unravelled, but studies of newts and salamanders should play a role in wider ecological studies that may ultimately be of economic importance.

There are a number of aspects of newt and salamander biology which have fascinated scientists for other reasons. The remarkable power of these animals to regenerate lost or damaged limbs has stimulated much active research over the years. Understanding the mechanism by which this is achieved could lead to important advances in medical research, and perhaps benefit those who suffer from deformed or damaged limbs as a result of injury. Recent research has also shown that salamanders produce skin secretions which are known to act as broad-based antibiotics. Perhaps these too may one day be developed for the treatment of human infections. Newts and salamanders are also favourite subjects among cell and molecular biologists. This is because their cells contain much larger amounts of DNA than most other animals. In addition, the structures within the cell that hold the genetic material, the chromosomes, are very large and easy to work with in microbiological studies.

Conservation of Species or Habitats?

From an ecological viewpoint it is important not to treat the conservation of newts and salamanders in isolation. Wildlife conservation involves the protection of species and the habitats in which they live. In an ideal world, protecting habitats should confer a high degree of security on the species they contain. All amphibian populations form an integral part of a wider ecological community, in which food supply, predators, diseases and habitats interact in a complex way. The goal of any conservation programme should therefore be to maintain whole ecological communities in which the interactions between species, as well as the species themselves, are maintained. Unfortunately habitat protection is logistically much more difficult to implement and enforce than is species protection, and the result is that many populations of animals and plants are still suffering declines in supposedly protected areas. In addition, a large proportion of conservation effort is funded by the public purse, and it is much easier to build a persuasive conservation campaign around a high-profile animal species than around a complex ecological system. In this regard newts and salamanders have a problem, as together with other ectothermic vertebrates, they have never commanded much charismatic appeal or sympathy with the general public. For those animals which have encountered ignorance and misconceptions throughout their history, conservation is as much about education and changing public attitudes as it is about preserving species and habitats. As far as newts and salamanders are concerned, there is still much to be achieved in this regard.

Threats to Newts and Salamanders

Global amphibian declines?

Amphibians are the only terrestrial vertebrates to have a skin that is fully permeable to water and a life cycle that includes both aquatic and terrestrial stages. Some conservationists argue that these two features make amphibians particularly vulnerable to environmental changes: a two-phased life cycle means that many species will be affected by changes to both aquatic and terrestrial habitats, and the permeable skin will render them particularly vulnerable to pollutants, acid rain, temperature extremes and ultraviolet light. To this end, certain amphibians may be good bioindicators of environmental change. Supporting such arguments is the fact that amphibian populations are declining in many parts of the world. Indeed, there have been some catastrophic losses of amphibian populations in north-west America, South America and Australia. What is more perplexing is that some of these declines are occurring in remote, pristine habitats where there is no straightforward explanation. So are amphibians providing a sort of early-warning system of a general global problem, or is the situation more complex than this?

What is certain is that not all parts of the world are affected in the same way, and that some species are declining faster than others. Asia, Africa and Europe seem less affected than the New World, at least in terms of the declines that cannot be readily explained by known causes. Also, there are far fewer reports of declines of newts and salamanders than there are of frogs and toads. This may merely reflect the fact that the anurans are more widespread and generally better studied, but even so, the inexplicable declines do not seem to have affected newts and salamanders in the same way as frogs and toads. Within Europe no amphibian species has become extinct within the last 100 years, although most species have shown declines over this period, mainly as a result of loss of habitat. Again, some species are more adaptable to environmental change than others. Smooth Newts, for example, have made something of a comeback in urban areas of England with the increase in the number of garden ponds, while the more specialised Crested Newt has not proved as adaptable.

In order to assign priorities for conservation action, the European reptiles and amphibians have been classified according to their vulnerability using the categories devised by the International Union for the Conservation of Nature, or IUCN (Corbett 1989). The assignment is made after taking into account such factors as geographical range, endemism, rate of decline, potential threats, and scientific interest of the species. Although no European newts or salamanders are considered to be threatened with extinction (i.e. in the 'Endangered' category), three types are considered to be 'Vulnerable'. These are the Blind Cave Salamander, the Golden Alpine Salamander and the Sardinian Cave Salamander (Table 7.1). These species could become threatened with extinction if action is not taken to address the threats that they face. Eight other species are considered to be less threatened, but still 'Rare' or 'Indeterminate'. In general, these threatened species are of concern because they have small, isolated ranges, and some of them occur only on islands. To some extent, then, their current plight is a consequence of historical factors which restricted their ranges to start with (see Chapter 3). However, when a species occupies a small geographical range it becomes much more vulnerable to other environmental changes caused by humans. Those environmental changes which have the biggest impact on newts and salamanders are now discussed below.

Habitat loss and fragmentation

The loss of habitat through human activities is undoubtedly the main threat to newt and salamander populations today. When the reasons underlying population declines are

TABLE 7.1 *Classification of European newts and salamanders according to the degree of threat (from IUCN Red List of Threatened Animals, 1994). Vulnerable: species not yet threatened with extinction (i.e. not yet 'Endangered'), but which could move into that category if causal factors for decline are not addressed. Rare: species not yet 'vulnerable', but which are at risk. Indeterminate: species known to be threatened but with insufficient data to place as 'Endangered', 'Vulnerable' or 'Rare'.*

Vulnerable	*Rare*	*Indeterminate*
Sardinian Cave Salamander (*Hydromantes genei*)	Monte Albo Cave Salamander (*Hydromantes flavus*)	Golden-striped Salamander (*Chioglossa lusitanica*)
Blind Cave Salamander (*Proteus anguinus*)	South-east Sardinian Cave Salamander (*Hydromantes imperialis*)	
Golden Alpine Salamander (*S. atra aurorae*)	Supramonte Cave Salamander (*Hydromantes supramontis*)	
	Sardinian Brook Salamander (*Euproctus platycephalus*)	
	Caucasian Salamander (*Mertensiella caucasica*)	
	Luschan's Salamander (*Mertensiella luschani*)	
	Lanza's Salamander (*Salamandra lanzai*)	

analysed, between 80% and 90% of causes are associated with habitat destruction. A recent survey in Britain highlighted urban development, agricultural improvement and pond drainage as the primary causes for the loss of amphibian habitat, with the loss of ponds considered to be more important than the loss of terrestrial habitat (Hilton-Brown & Oldham 1991). Similarly, in the countries which comprised the former Soviet Union, 11 out of a total of 13 species of newts and salamanders have been reported as declining. Deforestation and destruction of water bodies are cited as amongst the principle causes (Kuzmin 1994a).

Habitat destruction can affect wildlife in two ways. Firstly, it leads directly to smaller areas being available for wildlife. Small areas cannot support either the same number of animals or the same number of species as larger areas. Secondly, it leads to areas of land being increasingly divided up, or fragmented. This leads to populations becoming isolated from each other. Such 'insularisation' can lead to declines and extinctions independently of simple habitat loss. Small populations in isolated fragments are much more prone to extinction as a result of disasters, inbreeding, or other unpredictable factors (Figs 7.1 and 7.2).

The drainage, infilling, or overgrowth of a pond can therefore have two effects on newts and salamanders. In addition to providing one less area for feeding and breeding, it also removes a potential 'stepping stone' by which amphibians may colonise new areas. In Britain, estimates of pond loss over the past 30 years have varied between 15% and 25%. These figures could have been twice as high, had not the creation of new ponds, particularly in urban and suburban areas, offset the losses. Because of the stepping-stone effect, however, it is also necessary to take into account pond density. In Britain pond density varies between 0.4 and 30 ponds per km^2, with an average of around 1.4 (Swan & Oldham 1993).

Because the distances they can move on land are limited by moisture, newts seem to do best in areas of high pond density. In particular, Crested Newts seem to be absent from areas with a pond density of less than 0.7 per km², and Swan & Oldham recommend a target figure of at least 1 pond per km² in areas supporting this species.

In Denmark, the rate of pond loss has been estimated at about 19% over a 40-year period. Although this is a relatively low figure, a further 40% of breeding ponds suffered

some degree of alteration. Interestingly, the loss of amphibian populations over the same period (*c.* 50%) was actually higher than the rate of pond loss, emphasising the importance of habitat change rather than habitat loss (Fog 1995).

Habitat loss and fragmentation may mean that genes are lost from some populations along with the individuals that carry them. This may result in the species becoming less able to adapt to environmental pressures, or even worse, becoming inbred. Recent research has

FIG 7.1 *Habitat fragmentation. In an unfragmented habitat (left), ponds where amphibians have gone extinct may be recolonised by animals from neighbouring water bodies. In a fragmented habitat (right) barriers exist between ponds which prevent recolonisation following extinction. S. Elmhurst.*

FIG 7.2 *Urban development around a Crested Newt pond. This may destroy terrestrial habitat and isolate the breeding site from other water bodies (R. Griffiths).*

highlighted the importance of maintaining genetic variation within newt and salamander populations. A comparison of the DNA from different Crested Newt populations has shown that those from southern Europe are more genetically diverse than those in the north, and therefore deserve special consideration when it comes to conservation. The Southern Crested Newts (*Triturus karelinii*) in Turkey, for example, are genetically distinct from those in Bulgaria and the Balkans. The Northern Crested Newt (*Triturus cristatus*), on the other hand, is genetically much more uniform across its range (Wallis & Arntzen 1989).

Introduced predators or competitors

The ability of newts and salamanders to use ephemeral ponds and streams means that such habitats are usually free of predatory fish. Many amphibian species are vulnerable to fish predation at various stages of their life cycle, and in permanent water bodies the introduction of fish can have a dramatic effect on amphibian populations. In Britain about 36% of all ponds have fish, with sticklebacks and introduced goldfish being the most important species (Swan & Oldham 1993). In other parts of Europe the stocking of water bodies with salmonid fish for game purposes has either eliminated or severely reduced amphibian populations, including the Crested Newt in some areas. Although the toxic skin of the adult stage may provide a degree of protection against predators, Crested Newt larvae seem more vulnerable to fish predation than those of the smaller newt species. This is probably related to their habit of swimming within the water column in open areas, where they may be easily picked off by predators. At a pond in Sweden, the introduction of roach was followed by a decline in the Smooth Newt population and a shift in the age distribution so that older newts comprised a higher proportion of the population. This seemed to be caused by the fish preying on Smooth Newt larvae, which in turn resulted in fewer newts being recruited into the adult population (Aronsson & Stenson 1995). The introduction of trout has eliminated Crested Newts from many lakes in Norway, and sticklebacks, pike and perch are also detrimental to newts (Dolmen 1980). As many freshwater fish have diets which are similar to those of newts, competition for invertebrate food may also have a negative affect on newt populations.

Occasional draining of the pond should prevent the establishment of fish. If this is not possible, treatment of the pond with the chemical rotenone seems to be an effective way of eliminating such predators. Certainly, ponds in which fish are removed with rotenone may be rapidly colonised by newts soon after (McLee & Scaife 1993). This pesticide appears to have little long-term effect on invertebrates but the treatment should be carried out after amphibians have left the pond (Dolmen 1987).

Although illegal in Britain and other parts of Europe, in some areas non-native amphibians have been released either deliberately or accidentally into the wild. With one or two exceptions, such species do not appear to have expanded their ranges widely and therefore have had little, if any, impact on native populations of newts and salamanders. American Bullfrogs have been released in several areas, and if they succeed in breeding on a regular basis, their enormous tadpoles and voracious adults could have an impact on natural pond communities. Alpine Crested Newts (*Triturus carnifex*) were once widely exported for the pet and laboratory trade, and have been released into several places outside their natural range. They will certainly interbreed with, and possibly compete with, the other Crested Newt species, although there is no evidence that they have spread very widely from the points of introduction.

Other predators of newts and salamanders include ornamental ducks and geese. The presence of these animals on a pond may also affect the vegetation structure, and hence the availability of egg-laying sites. Other domestic animals, such as pet cats, may attack wild amphibians, although the skin toxins of newts and salamanders may afford them some degree of defence.

Collecting

Very little data are available on the threat posed to newt and salamander populations by collecting for the pet trade and laboratories. In the past, Fire Salamanders have been exported in large numbers from southern Europe to supply the demands of the pet trade and research institutes. Also popular have been the larger and more attractive species of newt, particularly Crested and Marbled Newts. Other species, such as the Alpine Salamander and brook salamanders, are difficult to keep in captivity without special conditions, and have never been in wide demand.

In more recent years, a greater awareness of conservation and animal welfare issues, coupled with stricter legislation, has led to a general decrease in the demand for wild-caught amphibians for laboratory work. However, newts and salamanders have always been much less popular than frogs in this respect and demand for scientific purposes has probably had a neglible impact on natural populations.

In Germany the overall number of amphibian species in the pet trade has also decreased over the past few years (Martens & Jelden 1992) and this may reflect a general trend in most of Europe. Indeed, with increased protection measures the proportion of indigenous newts and salamanders within the European-based trade has shown a downward trend over the past few years. However, this decrease may have been offset by an increase in imports of some Asiatic and North American newts (van Koolwijk 1987).

Political or economic upheaval can lead to a breakdown in the monitoring and regulation of collecting for the trade. According to Kuzmin (1994b), the break up of the former Soviet Union into independent republics has led to an upsurge in the collection of reptiles and amphibians in the pet trade, which is poorly regulated. Among the species affected are the Fire Salamander, Alpine Newt, Carpathian Newt, Banded Newt and Caucasian Salamander. In addition to recommending a tightening of measures to control wildlife trade, Kuzmin (1994b) also suggests that the development of captive-breeding programmes is needed to undermine commercial collection from the wild.

Small-scale collecting for private purposes may still be popular, and perhaps should be discouraged on welfare grounds. In most cases, though, such activities are unlikely to affect the local abundance of species. In Europe the collection of newts and salamanders for trade, private collections or scientific research needs careful monitoring and regulation, but at present should not detract from the more insidious threats posed by habitat destruction.

Pollution and acidification of breeding ponds

There is no doubt that rainfall in various parts of Europe has become more acid over the past 20 or 30 years. The effect of acidification on amphibians depends upon a wide range of factors. In some parts of Europe the acidification problem is much more severe than in others. Ponds and lakes in Scandinavia have suffered particularly alarming declines in pH, because air movements carry atmospheric pollution from the industrial areas of north-west Europe further north where it falls as acid rain. Small, ephemeral water bodies are particularly vulnerable to acidification as they may receive large quantities of unbuffered rainfall or snowmelt in the spring. This may cause short-term pulses of acidity which aquatic organisms are unable to tolerate. Other factors to take into account are the local soils and geology. Ponds situated on calcium-rich beds, such as limestone or chalk, will to some extent be buffered against acidic run-off as calcium salts within the soil will neutralise the acids before they can have a harmful effect. If ponds are situated on older rocks, such as granite and millstone grit, however, the buffering capacity of the soil will be much reduced and the effect of acidification will be more severe.

Amphibians vary widely in their tolerance to acidity. To some extent the degree of tolerance is related to habitat selection as species which habitually breed in boggy, soft-water pools are invariably more acid-tolerant than those which utilise limestone streams and pools. Tolerance also depends upon the phase of the life cycle. In general, eggs are more sensitive to acidity than are larvae, which in turn are more sensitive than adults.

Although the two species are very similar in general ecology, the Palmate Newt is more often found in soft-water ponds than the closely related Smooth Newt. Indeed, several studies have shown that Palmate Newts can be found in ponds with a pH as low as 4.0, whereas the Smooth Newt is rarely found below a pH of 6.0. What is rather surprising is that the eggs of both species survive equally well at low pH (4.5). Moreover, both species hatch earlier and at a smaller size at low pH, although Smooth Newts seem more affected as they hatch particularly early and are particularly stunted. In terms of larval development, the growth of both species is slowed by acid conditions but again there is no difference between the species. The negative effect of acidification on Smooth Newts seems confined to affecting size and time to hatching of eggs. In the field small size at hatching could mean that small Smooth Newt larvae are more vulnerable to predators than the larger Palmate larvae, and acidification may therefore exert its influence by altering the incidence of predation in favour of Palmate Newts in acid pools (Griffiths & de Wijer 1994, Griffiths *et al.* 1994).

Some potentially harmful metal ions may be present in freshwater either in very low concentrations or in a non-toxic form. Under acid conditions chemical reactions may occur which increase their toxicity to harmful levels. Aluminium, for example, is widely present in soils and freshwater in a non-toxic form, but becomes toxic under acid conditions. Newt larvae may become sluggish under such conditions and the gills become stunted. Acid rain and other pollutants may also indirectly affect amphibians by affecting their food supply. Water quality can dramatically affect the species and abundance of invertebrates present in streams and ponds, and this will have knock-on effects for the newts and salamanders which prey on them.

The impact of agricultural fertilisers on amphibians has been little studied. Nitrate fertilisers applied to fields can affect terrestrial-stage amphibians by direct contact, and aquatic amphibians may be affected by run-off into adjacent water bodies. In laboratory trials, Smooth Newt larvae raised under high levels of ammonium nitrate suffered from reduced size at metamorphosis. As feeding rates of the larvae were not markedly affected, the ammonium nitrate may cause reduced growth by a physiological mechanism, possibly involving an increase in water loss (Watt & Oldham 1995).

The overall impact of pesticides and pollutants on newt and salamander populations is difficult to assess. In the former Soviet Union, water pollution by cattle and industrial wastes was considered to be more harmful to amphibians than artificial fertilisers and pesticides (Kuzmin 1994a). Certainly, pollution is usually a less important factor than loss of habitat, although the two often go together and vary in importance between countries. In Norway, acid rain may represent one of the most serious threats to amphibians (Dolmen 1987). In Britain, however, only 6% of respondents to a survey considered pollution to be a contributory factor in the decline of local amphibian populations (Hilton-Brown & Oldham 1991).

Conservation Measures

Habitat preservation

If habitat preservation is to be effective, it must take into account the ecological requirements of the species that are to be conserved. This in turn requires a sound knowledge of the life history of all the species concerned. Providing an ideal pond for newts will not be an effective conservation measure if the surrounding terrestrial habitat is not capable of supporting the animals during their land-dwelling phase. Similarly, a mountain stream may provide an ideal breeding habitat for adult salamanders, but if short pulses of acid rain occur just after eggs are laid or larvae are deposited, successful development to metamorphosis may be impaired. A knowledge of the factors which influence the growth, development and survival of each stage of the life cycle is therefore needed before reliable decisions can be made regarding the designation of protected areas for newts and salamanders. Unfortunately, for the vast majority of species, such information is not readily available. The secretive, cryptic life-styles of newts and salamanders are not easy to study, and even in the best-researched species, such as the Smooth Newt and Fire Salamander, there are gaps in our knowledge of their precise requirements. This means that many decisions regarding the management and protection of habitats are based on 'educated guesses' about the requirements of the species concerned.

Protected areas

How large an area should be set aside for a newt or salamander population, and what features should be incorporated within it? Of course, the answer to this question will vary between species, but there are some general considerations that need to be addressed first.

Ideally, the population should be able to sustain itself within the limits of the protected area indefinitely into the future. The area must therefore be large enough to buffer the fluctuations in population size which seem to be a natural feature of many newt and salamander populations (see Chapter 6). Unless there is a single water body which is particularly large and stable, a population of a pond-breeding species is likely to need several water bodies of varying sizes and duration to provide such a buffer. Relying on a single pond within the protected area puts the population at risk from catastrophies which could result in the population being severely reduced or even made extinct. A single large pond which never dries out is risky for another reason. It provides an ideal habitat for many predatory fish which, sooner or later, could be introduced to the detriment of the amphibian population. In Germany, an analysis of amphibian communities which include Smooth Newts and Alpine Newts has shown that the chances of finding amphibians in a pond increase if the ponds are in groups. In other words, a pond in a cluster of five or more water bodies is more likely to contain newts than a single, isolated pond. As the total area of water available does not have an effect, this result cannot be attributed to the greater variety of microhabitats in groups of ponds (Mann *et al.* 1991).

If it is not possible to include more than one water body within a protected area, then an alternative strategy is to ensure that there are corridors of appropriate habitat linking the protected area to other water bodies nearby. Woodland seems to be particularly important for providing connections between amphibian breeding areas, while steep banks, fields and pastures are usually unfavourable (Laan & Verboom 1990). It is also obviously important that the water bodies are unpolluted, and can provide sufficient cover and food for the aquatic stages to behave, feed and grow normally (Table 7.2).

Much more difficult is assessing how large the terrestrial habitat should be and what landscape features should be incorporated within it. A pond system is rarely surrounded by a uniform area of terrestrial habitat, and the first problem to address is which areas in the vicinity of the ponds support most animals. This can sometimes be accomplished by setting up a series of drift fences and pitfall traps during the breeding migrations (e.g. Heyer *et al.* 1994). Migrating animals will be captured as they move towards the breeding areas, and by

TABLE 7.2 *Checklist of important features for a protected area for a newt or salamander population*

Aquatic habitat

(1) Several water bodies of variable size and permanence, or corridors linking a single water body with other areas
(2) Water body is unpolluted and of appropriate quality to allow all aquatic stages to grow and develop normally
(3) Water body is free of major amphibian predators (e.g. predatory fish), and provides sufficient food and microhabitats to enable larvae and adults to behave and feed normally

Terrestrial habitat

(1) Terrestrial habitat provides an appropriate microclimate and sufficient cover for all land-dwelling phases
(2) Terrestrial habitat allows animals to wander between water bodies, colonise other water bodies via habitat corridors, and is large enough for normal ranging behaviour
(3) Neither terrestrial nor aquatic habitats are subject to short-term disturbances which may influence any of the above

comparing counts from different parts of the area, the direction most animals are coming from can be determined. If, for example, over 75% of newts are captured while migrating from an easterly direction, this suggests that the habitat to the east of the ponds is more important than that elsewhere.

Once the important areas of terrestrial habitat have been identified, the next important question concerns how much of this habitat should be incorporated within the boundaries of the protected area. Obviously, there are many other economic and logistical factors which will influence this decision. However, it is important to take into account how far the species concerned ranges on land. Again, some data on this aspect may be gleaned from pitfall trapping newts or salamanders at different distances from the pond. Certainly, some species, such as Smooth Newts, seem to stay fairly close to their breeding ponds. Crested Newts seem to wander further afield, and have been found up to 1 km from the nearest breeding pond. Because we have little knowledge about the ecology and behaviour of newts and salamanders during the terrestrial phases, this is therefore a difficult point to address. A rule of thumb might be to determine the maximum distance that animals move from the breeding pond, and then include a band of land that distance wide around the edge that attracts most animals.

Many of the criteria described above can also be applied to newts and salamanders which breed in streams. Brooks and streams are linear habitats, and the area used either side may vary in extent according to the season. Outside influences within the catchment area are also important to consider. Water abstraction some miles from the stream, for example, may have severe repercussions for newts and salamanders which rely on certain levels of flowing water. As with ponds, a viable population may be dependent not on one but several streams. It is therefore important to ensure that individual streams are not isolated from each other by areas of inhospitable habitat.

Habitat creation and management

Ponds and streams are not stable habitats. If left protected, but unmanaged, they will change in character over time. This is due to the process of natural succession, which leads to one habitat type gradually replacing another. In the case of ponds, the process consists of the incursion of marginal plants, accompanied by a gradual silting up, so that less and less open water remains. Eventually the pond will become completely overgrown with vegetation, and succeed to marshland, and later on, to trees and scrub. At each stage of succession a new community of animals and plants takes over. This is a natural process, which in an ideal world would be offset by the steady creation of new water bodies as livestock watering holes are dug, or old mineral workings filled with water. Unfortunately, though, new ponds are not being created at the same rate that existing ponds are succeeding to dry land. Simple neglect of ponds is therefore a major factor contributing to loss of breeding habitat. This means that existing ponds must be managed to maintain a variety of microhabitats which will support as diverse an ecological community as possible.

In general, newts and salamanders prefer ponds at an intermediate state of succession, with a variety of vegetation types to provide hiding places and egg-laying sites, but not so choked that there is no open water. Management of small water bodies therefore involves periodically removing submerged or encroaching vegetation so that succession is kept under control and some open areas of water and shoreline are maintained (Fig. 7.3). South-facing shorelines should generally be cleared of all trees and exposed to the sun. This will help provide a temperature gradient within the pond and enable the animals to select the

FIG 7.3 *Removal of encroaching reeds from a newt breeding pond. Such management may be necessary to maintain healthy populations (R. Griffiths).*

most appropriate temperatures for egg and larval development. Of course, timing of management activities is crucial, as it should not overlap with breeding times or other periods when eggs, larvae or adults are likely to be present in the pond in large numbers. The best strategy is to organise management when breeding is finished and most larvae have metamorphosed, which is often the winter months. If the management is organised on a cyclical basis, with say, a quarter of the pond being managed each year for 4 years, then any negative effects on overwintering amphibians or other pond life will be minimised. In Denmark, the restoration of old breeding ponds has been highly successful in increasing the number of Alpine Newt populations from 46 in 1978 to 156 by 1994 (Fog 1995).

The creation of ponds in gardens has proved an effective conservation measure for some species (Fig. 7.4). One survey in an urban area of north-east England revealed that just over 6% of gardens had ponds, and 40% of these contained breeding populations of amphibians (Banks & Laverick 1986). Out of the newts, the Smooth Newt seems to do best in garden ponds. Crested Newts are more difficult to establish, but their success in garden ponds probably depends on whether other viable populations exist in the area. The author's own garden pond was colonised naturally by all three native newt species in the first year of its construction. The Crested Newt colonisers consisted of just two individuals – a subadult and a female. As the female laid fertile eggs, and no males were observed in the pond, she presumably had mated in a nearby pond. The most likely source of the colonisers was a large garden pond about 300 m away.

Managing the terrestrial habitat is also important. Providing adequate cover in the form

FIG 7.4 *Garden ponds can provide important breeding areas for newts. This pond in Wales supports Crested, Smooth and Palmate Newts (R. Griffiths).*

of scrub, long grass, log piles, stones etc. is important for most species of newts and salamanders. Rock piles may be needed for those salamanders that inhabit narrow cracks and crevices. Short-grass areas should be allowed to succeed to rough grassland or scrub cover, but pure coniferous woodland should be discouraged, as it cannot support sufficient cover at ground level. Minimising other types of disturbance to the area is also a crucial part of site management. The general public can be discouraged from entering sensitive parts of the site by providing well-marked footpaths through less sensitive areas. Usually managing water bodies and the surrounding habitat in this way will also benefit many other species too.

Translocations

By now it should be clear that the creation of new water bodies in appropriate areas of terrestrial habitat is one of the most effective conservation measures that can be carried out for newts and salamanders. Ideally, newly created ponds should be left to be colonised naturally. Providing there are corridors of suitable habitat linking the protected area with other amphibian breeding sites, this can often be achieved rapidly. Newts will colonise new ponds soon after they have been created, although it may take a few years for reasonable numbers to build up. However, there may be circumstances when the deliberate release, or *translocation*, of animals into the areas of suitable habitat needs to be considered. There are several different types of translocation. An *introduction* is where a species is released into an area where it has not previously been known to occur. There need to be very sound conservation reasons for releases outside a species' range, as there may be good ecological reasons why it does not occur there in the first place. More usually, a translocation takes the form of a *reintroduction*, where a species is released back into an area that it formerly occupied, but from which it has recently disappeared. *Restocking* is where animals are released to supplement an existing population, perhaps one that has reached a low ebb from which it is unlikely to recover. Again, there need to be good reasons for such a translocation, as there are risks involved with overcrowding existing populations.

Planning a translocation

Before carrying out any translocation, there are a number of important factors that need to be considered. Firstly, a thorough ecological study of the new area needs to be conducted to ensure that it is suitable for the species concerned. If the species occurred in the area some time before and went extinct, then a full appraisal of the factors that led to that extinction needs to be carried out.

Secondly, the population which is to provide the donor animals needs to be considered. Is it large enough to withstand the abstraction of individuals, and is the taking of these animals likely to have any knock-on effects on other animals or plants? These are often difficult questions to answer, and often the only justification for removing animals from a site is if it is going to be destroyed.

Thirdly, a decision must be made regarding the stage of development at which the donor animals are to be taken. Adult newts and salamanders have strong homing instincts (see Chapter 4), and can become disorientated if removed from their native pond. This may result in them wandering off if placed in a new pond some distance away. The translocation of eggs or larvae may therefore be more effective, but large numbers of these stages will need to be used to offset the very high natural mortality that occurs naturally prior to metamorphosis. Probably the best strategy is to translocate well-developed larvae which have a good chance of survival to metamorphosis.

Fourthly, the genetic constitution of the donor animals must be considered. To some extent all animals will have become adapted to the local conditions in which they live and reproduce. If they are to be translocated elsewhere, it is important to consider whether these adaptations, acquired through many generations of natural selection, will adequately equip them for life in their new environment. In southern Europe, newts breed in the autumn and winter months, which are much milder than those further north. If a number of these animals were translocated to ponds in northern Europe, it is quite possible that they would attempt to breed at the same time of year that they do in the south. This would result in reproduction taking place at the most inhospitable time of the year, and the population would be unlikely to survive in the long term. If a translocation is to be attempted, then, a local population should be used to provide donor animals wherever possible.

Lastly, it is important to minimise the risk of transmitting diseases or unfavourable organisms in a translocation. The movement of newts or salamanders from one site to another may inadvertently result in pathogens being transmitted between populations. If the newly created area is totally isolated from other populations then this is unlikely to be a problem. However, there are certainly other organisms which could be moved to the new pond along with the translocated animals, and which should be avoided. These include various 'pest' plants, such as duckweed (*Lemna*) which can carpet the surface of a pond and cut out the light, or water stonecrop (*Crassula helmsii*), a non-native species which will rapidly choke the water column.

Captive-breeding

The release of captive-bred animals into new areas circumvents some of the problems involved with translocations. Apart from the original animals which comprised the founder stock for the captive colony, such an exercise will not require the abstraction of animals from natural populations. Moreover, newts and salamanders are in many ways ideal subjects for breeding and release programmes. Many species breed well in captivity, and their high fecundity means that very large numbers of animals can be raised quickly and cheaply (in captivity eggs and larvae are not faced with the array of predators that deplete wild populations). Very often captive-bred animals are in better condition and have lower parasite loads than wild animals.

Some problems may remain, however. Baker (1994) compared the fate of captive-reared Crested Newts released into the wild with those from a natural population. Over a 3-year period, fewer captive-reared newts were recaptured in the population than wild animals. There are several possible reasons for this. It is possible that the captive-rearing regime had some adverse affect on the newts, which meant that they survived less well when released. Certainly, those animals which were released were smaller than most of the other wild animals. Another possibility is that captive rearing allowed some, possibly weaker, newts to survive which would have been selected out by predation or other factors in nature. When released into the wild such animals were unsuited for survival in the original habitat and quickly succumbed to natural pressures. Lastly, captive rearing may not have allowed newts to learn where the most appropriate terrestrial habitat was located. Consequently, they became disorientated when they left the pond and did not return. These areas require much further research, but the safest strategy at present is to release captive-bred newts and salamanders only into areas within their natural range but where direct mixing with wild animals is unlikely.

Follow-up monitoring of translocations

Few studies have attempted to assess the success of translocations for newts or salamanders. In Britain, 86 Crested Newt translocations were carried out between 1970 and 1990, but under half of these were followed by monitoring of the animals that were released. Out of the sites that were monitored, newts were not recorded at 14, a failure rate of 33%. In most cases the failure of the translocation was a result of predators or unsuitable habitat. Indeed, most of the successful translocations were to garden pond habitats which were probably specifically managed for the newts (Oldham *et al.* 1991).

In sum, the idea of creating new populations of newts and salamanders by translocating animals from one area to another is intrinsically appealing, but in practice may be difficult to carry out effectively and safely. As a conservation measure, translocations should be a last resort, only to be used when other efforts to safeguard the species and its habitat have failed. Translocations are most often carried out when large numbers of animals are collected in rescue operations from sites about to be destroyed. Such operations may prolong the lives of amphibians which would otherwise be doomed, but in terms of conservation of viable populations their benefits are questionable; from an ecological viewpoint what is good for the individual animal may not be good for the population as a whole. A common problem is how the rescued animals are to be disposed of. Releasing them into an existing population may cause overcrowding, or genetic or disease problems as described above. Even worse, there may simply be no suitable habitat available locally. Follow-up survey work should be a formal requirement for all translocations, but because of the natural fluctuations that occur in newt and salamander populations, is likely to be long term. Any translocation programme therefore requires very careful planning and execution, and should take account of all necessary legislation relating to the conservation, welfare and movement of animals.

LEGISLATION

Within Europe there is a bewildering array of laws, conventions, appendices and articles relating to environmental protection. National and international legislation varies considerably in its effectiveness in protecting wildlife, and there exists an almost continuous process of review and updating which means that any summary of legislation relating to particular groups of animals and plants will quickly become out of date.

The Bern Convention on the Conservation of European Wildlife and Natural Habitats provides a framework for the protection of animals and plants within the area covered by the Council of Europe. Appendices to the Convention list the species which it covers. Amphibians are listed in Appendices II and III, which schedule species as strictly protected, or in need of regulation where exploitation is taking place, to prevent local declines, respectively (Table 7.3). A committee of amphibian and reptile experts provides advice to the

TABLE 7.3 *Newts and salamanders listed in Annex II of the Bern Convention. *indicates those also listed in Annex IIa of the Habitats Directive*

Golden-striped Salamander (*Chioglossa lusitanica*)*
Pyrenean Brook Salamander (*Euproctus asper*)
Corsican Brook Salamander (*Euproctus montanus*)
Sardinian Brook Salamander (*Euproctus platycephalus*)*
Monte Albo Cave Salamander (*Hydromantes flavus*)*
Sardinian Cave Salamander (*Hydromantes genei*)*
South-east Sardinian Cave Salamander (*Hydromantes imperialis*)*
Italian Cave Salamander (*Hydromantes italicus*)
Supramonte Cave Salamander (*Hydromantes supramontis*)*
Luschan's Salamander (*Mertensiella luschani*)*
Blind Cave Salamander (*Proteus anguinus*)*
Alpine Salamander (*Salamandra atra*)
Golden Alpine Salamander (*Salamandra atra aurorae*)*
Spectacled Salamander (*Salamandrina terdigitata*)*
Alpine Crested Newt (*Triturus carnifex*)
Northern Crested Newt (*Triturus cristatus*)*
Danube Crested Newt (*Triturus dobrogicus*)
Southern Crested Newt (*Triturus karelinii*)
Italian Newt (*Triturus italicus*)
Carpathian Newt (*Triturus montandoni*)

Convention on herpetological matters, and parties to the Convention meet on an annual basis to discuss matters of implementation across Europe. Those countries which have ratified the Bern Convention are obliged to implement national legislation for the protection of those species listed in the appendices. For example, the enactment of the Wildlife and Countryside Act (1981) is intended to meet Britain's obligations under the Bern Convention. However, there is considerable variation between countries in how they have interpreted their obligations under the Bern Convention, and consequently the effectiveness of the legislation that has been implemented varies across Europe. Although many European countries have laws which make the collection, killing or trading of newts and salamanders illegal, legislation protecting amphibian habitats is generally much weaker. In some countries, then, netting a Crested Newt from a pond would be regarded as an offence, whereas filling in the pond itself would not. Other loopholes have sometimes allowed other laws to overrule or circumvent those covering wildlife protection. Even where national legislation may be sound, enforcement may be weak. In countries where water bodies reach a density of more than 10 per km^2, resources may not be available to monitor pond losses and prosecute those who deliberately destroy sites holding protected species. A further problem is that in many countries the implementation of national legislation is delegated to local regions. Variation between regions in conservation expertise and resourcing may once again lead to inconsistent enforcement of wildlife protection laws within a single country. The implementation of 'Pond Protection Orders' would therefore seem to be a potentially powerful conservation tool, but such measures have yet to be widely applied in Europe. Denmark is probably unusual in having every pond of 100 m^2 or more protected by law (Fog 1995).

More recently the European Union Council Directive on the Conservation of Natural Habitats and of Wild Fauna and Flora (frequently abbreviated to the simpler 'Habitats Directive') has provided a further framework for the enactment of conservation legislation within Europe. Under the Directive member states are required to implement legislation to protect certain species and habitats which are a conservation priority in their region. In some countries, including Britain, the national legislation already in place goes some way to meeting these requirements, although some further measures may be needed for certain species. The Crested Newt (*Triturus cristatus*), for example, is listed in Annex IIa of the Habitats Directive, and requires that areas of land be designated as 'Special Areas of Conservation' (or 'SACs') for this species. Research is currently under way to identify such areas and implement the necessary protection.

Signatories to the Washington Convention on International Trade in Endangered Species (1973), also known as CITES, are obliged to ban or regulate the trade in species which are listed in the appendices of the convention. Seventy-two types of amphibian are included under CITES (1.7% of all known species), but these do not include any European newts or salamanders.

In many countries there are additional laws which govern issues related to animal welfare, pets, zoos, scientific experiments, pollution, and public health and safety. Such laws may have some bearing on newts and salamanders if these animals are being used for activities which involve such issues.

CONCLUSION

The problems facing newts and salamanders are not unique. They are basically the same problems which are threatening all species of animals and plants the world over. All groups of living organisms have evolved differences in physiology and behaviour which will mean that they respond to environmental change in different ways. Like many other animals, newts and salamanders have qualities which make them sensitive indicators of environmental changes. On the down side, however, their small size, secretive way of life and population fluctuations make them less convenient to use in this regard. Amphibians have been around for some 360 million years, and the ancestors of present-day newts and salamanders saw periods of great climatic upheaval and the rise and fall of the dinosaurs. In order to ensure that they endure for a little longer, we need to understand how adaptable our present-day species are to a changing climate, as well as making sure they have somewhere to live in the future.

CHAPTER 8

Species Accounts

Triturus alpestris life cycle (not to scale)

AS new discoveries are made about the evolutionary relationships between species, ideas change about how they should be classified. In addition, biologists cannot always agree on whether a particular animal warrants classification as a full species in its own right, or is just a subspecies of another species. Newts and salamanders are no exception in this regard, and the species accounts listed below will no doubt need to be revised in the future as new research comes to light. Both scientific and common names are provided for each species. However, it is important to bear in mind that some species do not have widely used common names, or may be known by different names in different parts of their range. The species descriptions focus on those features which will most readily assist identification in the field and should be used in conjunction with the colour plates. Differences between subspecies may be very subtle, and some forms can only be reliably distinguished using biochemical techniques. In such cases the location where the animal was found may be helpful in arriving at its identity. The eggs and larvae of several species are also difficult to distinguish in the field, and again, the place where these stages were found may provide a guide to their identity.

The references listed for each species have deliberately focused on more recent works. These should provide an entry to older, and sometimes more detailed literature, if the reader is interested in further information on classification, distribution or general ecology. For more general information on the identification of European newts and salamanders, the reader is referred to one of the standard field guides, such as Thorn (1968), Arnold & Burton (1978), Ballasina (1984) or Nöllert & Nöllert (1992).

FAMILY: PLETHODONTIDAE

This family of lungless salamanders has undergone a vast radiation in the New World, but is currently represented in Europe by a single genus and seven species (Lanza & Corti 1993). The most appropriate generic name for the European species is disputed. *Speleomantes* has been preferred by some European authors since the 1980s, although *Hydromantes* has a longer and more consistent usage. At the time of writing (December 1995), a decision is awaited from the International Commission on Zoological Nomenclature regarding which name has priority. All of the species are similar in general appearance, and are most reliably distinguished using biochemical techniques. They are characterised by a slender body with a cylindrical tail which accounts for just under half the total length. The head is broad and rounded and has large eyes. The tongue is protrusible, and can be everted to a distance of about 40 mm to catch prey, further than any other European salamanders. The toes are stubby and partially webbed on the hind feet. The sexes are similar in appearance, but males have an oval-shaped glandular swelling on the chin (the mental gland). A pair of small sensory tentacles may also be visible on the upper lip. These are associated with a shallow groove running from the nostril to the lip (the nasolabial groove). The natural history of the European species is poorly understood.

Hydromantes ambrosii Lanza, 1955

Spezia Cave Salamander (Plate 1)

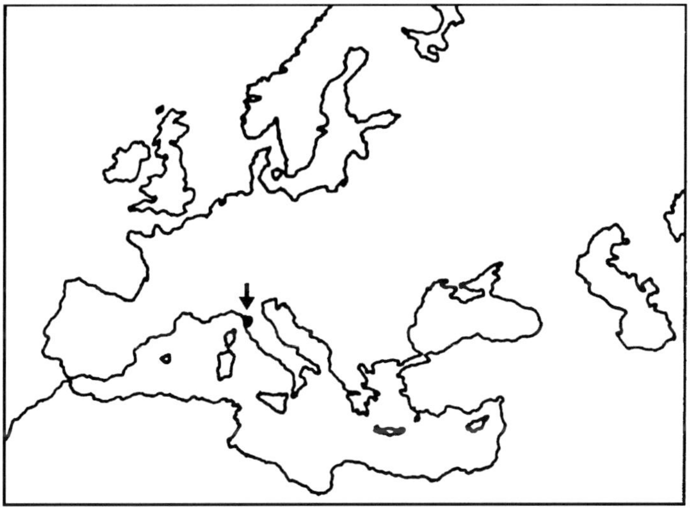

Distribution: La Spezia province in north-west Italy, east of La Spezia town.
Description: Up to 13 cm, females slightly larger than males. Colouration varies, but the body is usually brown with lighter dappling along the flanks. There may also be darker markings along the back.
Habitat: Shady, upland areas in the vicinity of streams and limestone valleys. Lives in rocky outcrops or within leaf litter along the stream valleys, or in natural and artificial caves and tunnels.
Habits: Entirely terrestrial salamander which spends much of its life in hiding places. In surface habitats (i.e. in the vicinity of surface streams) activity seems to be related to rainfall with most surface activity occurring in spring and autumn. In subterranean habitats seasonal activity is more related to temperature and peaks during the summer months. Males, females and juveniles may use slightly different areas of the same cavern or tunnel, with juveniles staying close to the entrance. However, all three stages may move slightly deeper underground during the summer months. Foraging occurs on the rock face walls, and irregular surfaces with many cracks and crevices are used more frequently than smooth cave surfaces. Breeding and egg brooding takes place within deeper crevices and is difficult to observe, but mating probably takes place in the spring or early summer with egg deposition in the autumn or winter. Population structure consists of about three subadult cohorts (i.e. 1, 2 and 3 years old) together with adult males and females which mature at 4 and 5 years, respectively.
Eggs and larvae: Small clutches of 5–12 eggs are laid in the winter months and brooded by the female. These undergo direct development with fully metamorphosed juvenile salamanders emerging from the eggs. Females sometimes cannibalise their eggs.
References: Lanza & Corti (1993); Salvidio (1993); Salvidio *et al.* (1994).

Hydromantes strinatii Aellen, 1958

North-west Italian Cave Salamander (Plate 2)

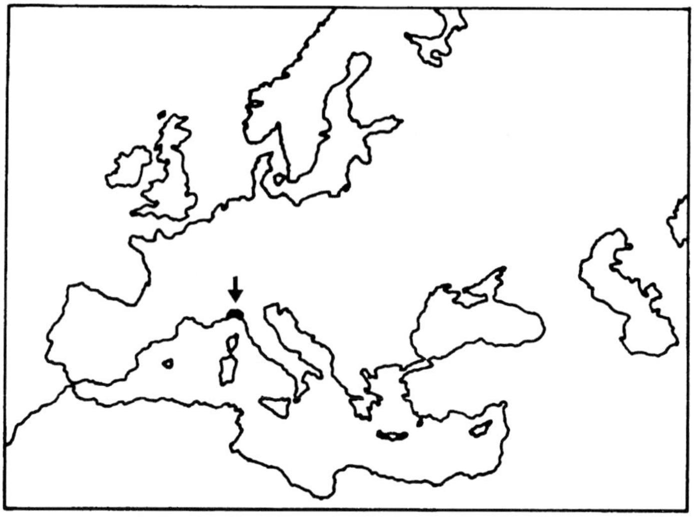

Distribution: South-east France and north-west Italy in Liguria, west of La Spezia town.
Subspecies:

Hydromantes strinatii argentatus: North-west Italy in the Varatella valley, in west Liguria.

Hydromantes strinatii bonzanoi: North-west Italy in the Prino, Impero, San Pietro and Arroscia valleys, in west Liguria.

Hydromantes strinatii ligusticus: North-west Italy between Genoa and Rapallo in central Liguria.

Hydromantes strinatii strinatii: North-west Italy in west Liguria and south-east France in the Maritime Alps and Haute-Provence Alps.

Description: Up to 13 cm, females slightly larger than males. Colouration is very variable, and the body may be brown, reddish or grey–green with lighter dappling along the flanks.
Habitat: In the vicinity of streams and seepages in mountain regions (often limestone areas), where it inhabits rocky outcrops, caves, crevices or surface debris.
Habits, eggs and larvae: As in *H. ambrosii.*
References: Lanza *et al.* (1986); Nardi (1991); Lanza & Corti (1993).

Hydromantes italicus Dunn, 1923

Italian Cave Salamander (Plate 3)

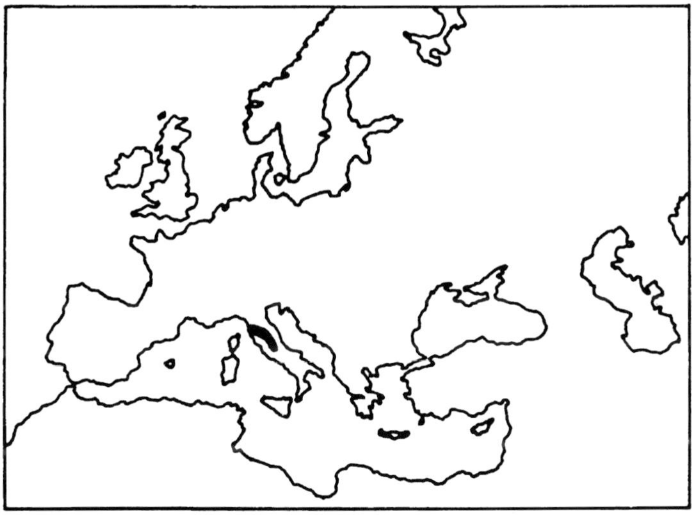

Distribution: North-east and central Italy.

Subspecies:

Hydromantes italicus italicus: From the Umbro Marchigiano Appenines north-west to the Tosco Emiliano Appenines.

Hydromantes italicus gormani: In the east part of the province of Carrara and the Appenines in the province of Modena.

Description: Up to 12 cm. Upper surface yellow, reddish or brown with lighter coloured mottling. Underside greyish with white speckling.

Habitat: In the vicinity of streams and seepages in mountain regions (often limestone areas), where it inhabits rocky outcrops, caves, crevices or surface debris.

Habits, eggs and larvae: As in *H. ambrosii.*

References: Lanza *et al.* (1986); Nardi (1991); Lanza & Corti (1993).

Hydromantes genei (Temminck & Schlegel, 1838)

Sardinian Cave Salamander (Plate 4)

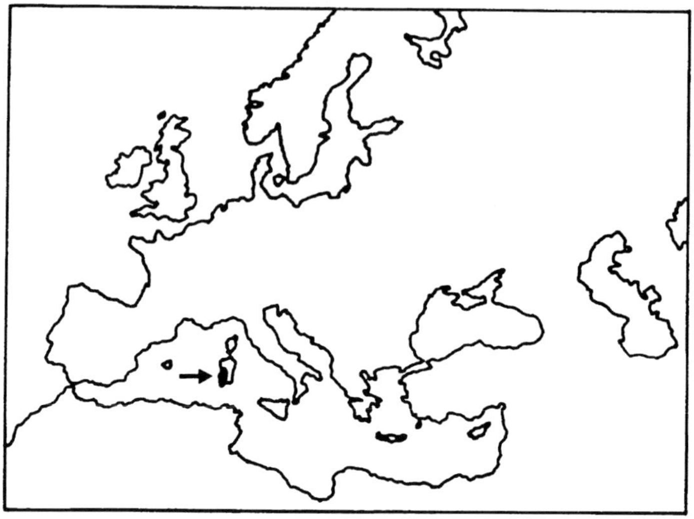

Distribution: South-west Sardinia.
Description: A slender salamander up to 13.5 cm. Colour varies from yellowish-brown to dark brown with lighter mottling. Belly is whitish, sometimes with dark speckling.
Habitat: Damp, mountain regions, where it inhabits rocky outcrops, caves, crevices or surface debris at around 850 m.
Habits: Fully terrestrial, secretive and exclusively nocturnal. Behaviour of this lungless salamander is related to surface moisture. It forages on rock surfaces and amongst surface litter on damp nights.
Eggs and larvae: Undergoes direct development on land. Eggs are laid in cracks and crevices and hatch into fully developed young salamanders.
References: Lanza *et al.* (1986); Nardi (1991); Lanza & Corti (1993).

Hydromantes imperialis Stefani, 1969

South-east Sardinian Cave Salamander (Plate 5)

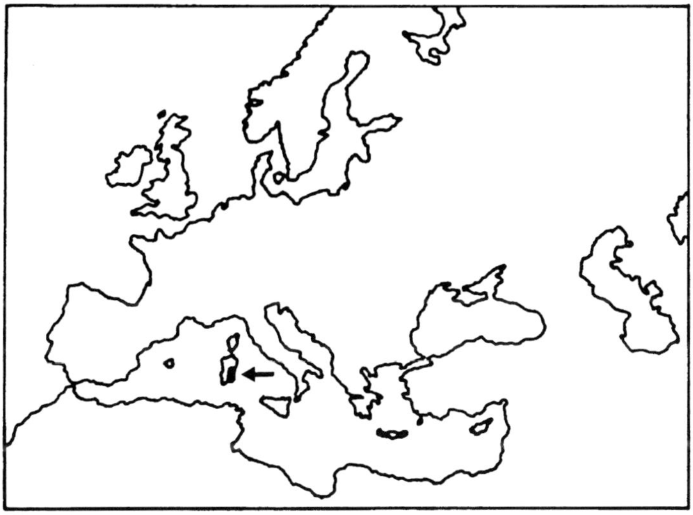

Distribution: South-east Sardinia.
Description: Both sexes up to 12.5 cm. The upper surface is purple–brown, with scattered golden spots with a metallic sheen. Some individuals may have a lighter brown mottling on the back. Only *Hydromantes* which emits a strong smell.
Habitat: Damp, mountain regions, where it inhabits rocky outcrops, caves, crevices or surface debris.
Habits, eggs and larvae: As in *H. genei.*
References: Lanza *et al.* (1986); Nardi (1991); Lanza & Corti (1993).

Hydromantes supramontis Lanza, Nascetti & Bullini, 1986

Supramonte Cave Salamander (Plate 6)

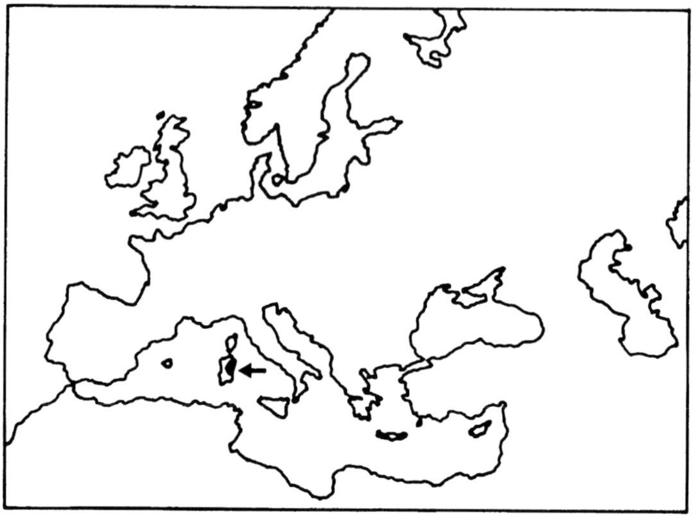

Distribution: Central eastern Sardinia.
Description: A large species, up to 13.5 cm; tail slightly shorter than the body. Slightiy narrower head than *H. flavus*. Colouration is variable. Dorsal surface usually darkly coloured with yellow, grey or olive-green mottling. Young specimens tend to be greenish with a metallic sheen. Belly and throat whitish, usually with a number of dark spots (cf. *H. flavus*).
Habitat: Damp, mountain regions, where it inhabits rocky outcrops, caves, crevices or surface debris.
Habits, eggs and larvae: As in *H. genei*.
References: Lanza *et al.* (1986); Nardi (1991); Lanza & Corti (1993).

Hydromantes flavus Stefani, 1969

Monte Albo Cave Salamander (Plate 7)

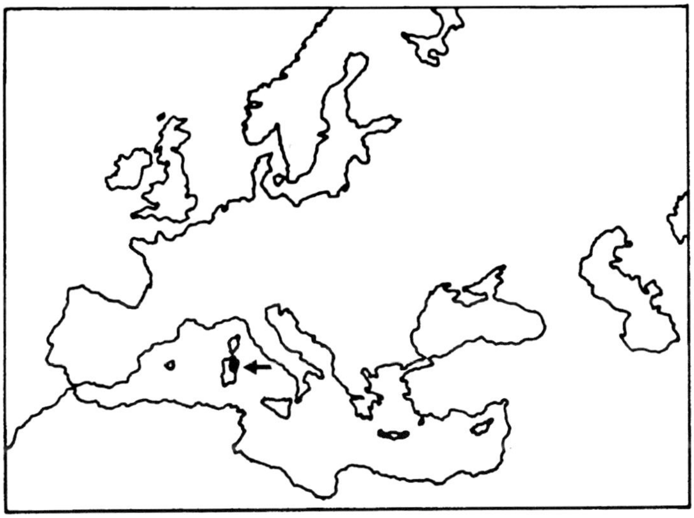

Distribution: North-east Sardinia.
Description: Largest *Hydromantes*, up to 15 cm. Reaches sexual maturity at 7–9 cm. Head wider than *H. supramontis*. Background colour varies between specimens from yellow, grey, green, brown through to black. Underside usually without spots.
Habitat: Damp, mountain regions, where it inhabits rocky outcrops, caves, crevices or surface debris, often in association with dense vegetation.
Habits: Fully terrestrial, secretive and exclusively nocturnal. Forages on rock surfaces and amongst surface litter on damp nights. Population size in one cave system was recorded at a minimum of 100 individuals, with females outnumbering males by about 2:1 The density of salamanders in an adjacent forest site was at least 300 per ha. Adult salamanders may cannibalise young salamanders and eggs.
Eggs and larvae: Undergoes direct development on land. Eggs are laid in cracks and crevices and hatch into fully developed young salamanders 2–2.5 cm long.
References: Lanza *et al.* (1986); Voesenek *et al.* (1987); Nardi (1991).

FAMILY: SALAMANDRIDAE

This family includes those amphibians which may be regarded as 'typical' newts and salamanders. It is the most diverse family in Europe, and includes several species which are widespread and common across their geographical ranges. The family includes both large- and small-bodied species. The tail may be cylindrical or ovoid in shape (salamanders) or flattened from side to side (newts). It is usually just under half the body length, although three genera (*Chioglossa*, *Mertensiella* and *Salamandrina*) are characterised by species with longer tails. In those species belonging to the crested newt 'species group', the Wolterstorff Index may assist in identification. This index is calculated as the length of the front legs divided by the distance between the front and hind legs (see Arntzen & Wallis (1994) for a review of the method). The species descriptions for newts refer to animals in breeding condition. This is when males develop conspicuous breeding characters and different sexes, as well as different species, are quite different in appearance. At this time of the year, females may have a plump appearance as they will be carrying eggs. Outside the breeding period newts lose their breeding characters, their skin adopts a drab, velvety texture, and they may be more difficult to identify. Sexual differences are less marked in most of the salamanders, but breeding males usually have a swollen cloaca. Immature newts or salamanders lack the breeding characters at all times of the year, and usually resemble small females.

Triturus boscai (Lataste, 1879)

Bosca's Newt (Plate 8)

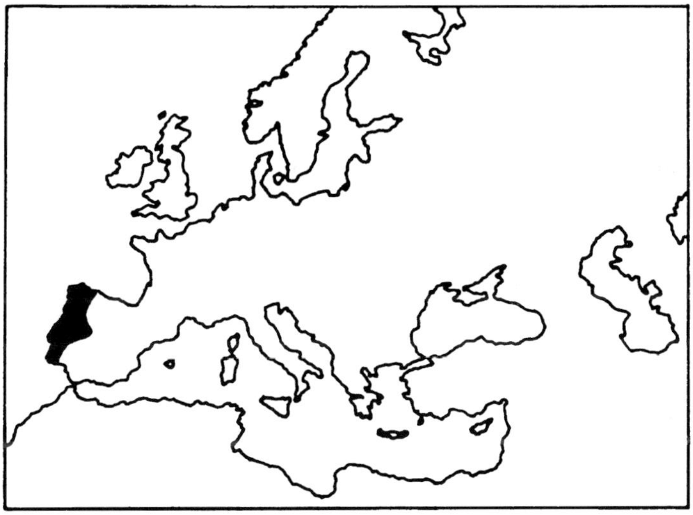

Distribution: Central and western Iberian peninsula.
Description: A small newt (up to 9 cm), males slightly smaller than females. Both sexes superficially similar to Palmate or Smooth Newts, but sometimes have a yellow dorsal stripe running from the neck to the tail. In males the breeding dress is much less elaborate than in most other species of newts. They do, however, have ridges along each side of the back (which gives the newt a square-backed appearance), and a low, smooth-edged crest on the tail. In addition, there may be a short, whitish protuberance at the tail tip. A further feature which distinguishes it from similar species is the absence of a dark stripe through the eye. The upper parts are yellow-brown or olive-brown, with dark and irregularly distributed spots. The belly is orange to bright red with whitish margins bordering the flanks. The orange colouration extends along the lower margin of the tail. The throat is paler than the belly, but both may be spotted. The female is similar to, but slightly darker than the male, and lacks the ridges on the back and crest on the tail. She also has slightly fewer spots, and a slightly pointed cloacal swelling.
Habitat: Found in cool, clear ponds or moving water, usually in hilly areas. Occurs in streams more often than other species of newts.
Habits: Little studied in nature, though probably has a typical newt life cycle similar to that of the Smooth Newt. Breeding takes place between December and March, and newts leave water by June to spend the rest of the year on land. Unlike other newts there may be bodily contact during courtship, with the female nudging the male with her nose. By undulating the whitish tail tip in front of the female, the male mimics the movements of a possible prey item, and in this way attracts a potential mate.
Eggs and larvae: Like other newts, the eggs are laid individually on water plants, and hatch into pond-type larvae which are difficult to distinguish from other newt species. Larvae grow up to 45 mm before metamorphosis, and have a rather blunt or rounded tip to the tail.
References: Garcia-Paris (1985); Rafinski & Pecio (1992); Faria (1993, 1995).

Triturus helveticus (Razoumowsky, 1789)

Palmate Newt (Plate 9)

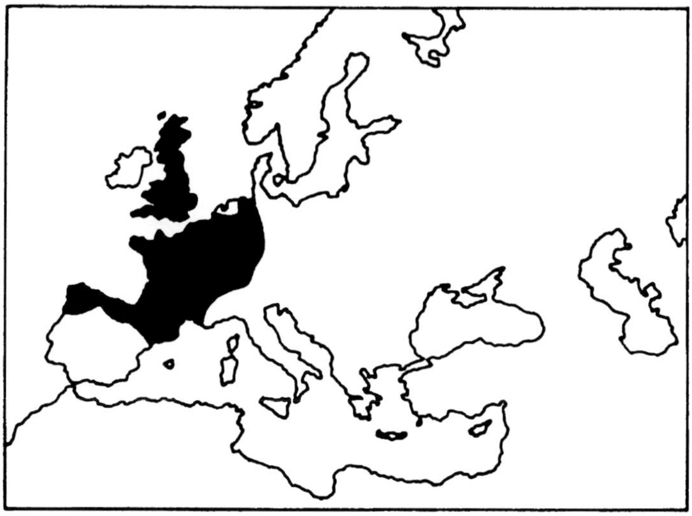

Distribution: Western Europe, ranging from Britain and northern Germany, through most of France, to northern Spain and Portugal.

Subspecies:

Triturus helveticus helveticus: All of the above range apart from north-west Iberia.

Triturus helveticus punctillatus: Sierra de la Demanda, Spain.

Triturus helveticus alonsoi (= *T. h. sequeirai*): Northern Portugal and north-west Spain.

Description: Smaller than the Smooth Newt (up to 9.5cm), with males smaller than females. Individuals from the southern part of the range are often smaller than those from further north, but the validity of the Iberian subspecies is questionable. Smooth, olive-green or brown background colour; males with dark spots on sides and two rows of spots on the tail bordering a pale orange flash. Breeding males develop a low, smooth-edged crest along the back and tail. They also have black webbing on the hind feet, and a fine filament protruding from the end of the tail. A short, narrow eye-stripe runs along each side of the head. Two ridges run along each side of the back giving the body a rather square appearance. Outside the breeding period many of these characters are lost and males are similar to females. The two sexes can always be distinguished, though, by inspection of the cloaca, which is always bulbous and darkly pigmented in male newts. The female lacks any webbing, tail filament or ridges along the back. She is usually of uniform colouration, although a variable number of small dark spots occur on the body and tail. Both sexes have a pale orange or yellow belly, sometimes with a few dark spots. The throat, which is always pinkish and unspotted, is a useful feature for distinguishing between Palmate and Smooth Newts. Immature newts may be identified as Palmate Newts by a light-coloured dorsal stripe which runs from the neck, down the back and on to the tail. This is yellow or yellow–orange in colour and of equal intensity along its length.

Habitat: Although it occurs up to 2000 m in some places and in Britain was once considered to be an upland species, the Palmate Newt actually occurs in many lowland ponds as well. It seems to occur more frequently in soft-water areas than other newts, but its tolerance of acidity is only marginally better than that of the Smooth Newt. The association between Palmate Newts and soft water may therefore be due to other factors, such as the low nutrient status or metal ion concentration of such ponds. It may also have a preference for the cooler, shadier ponds which are often adjacent to woodlands.
Habits: The annual cycle and behaviour of the Palmate Newt are very similar to those of the more widespread Smooth Newt. Indeed, where their ranges overlap the two species are often found in the same pond. In northern Europe Palmate Newts enter water in February or March and stay in water until June or July. There is a two-peaked daily activity pattern, with most activity occurring in early morning or just after dusk. During the daytime newts hide amongst dead leaves on the bottom of the pond or within vegetation.
Eggs and larvae: The eggs are laid individually on water plants, and both eggs and larvae are indistinguishable from those of the Smooth Newt in the field. However, the fully grown larva reaches about 40 mm, which is slightly smaller than the Smooth Newt. The pattern of egg-laying and development is also the same, although Palmate Newts seem to metamorphose slightly earlier.
References: van Gelder (1973); De Fonseca & Joqué (1982); Veith & Dörr (1985).

Triturus italicus (Peracca, 1898)

Italian Newt (Plate 10)

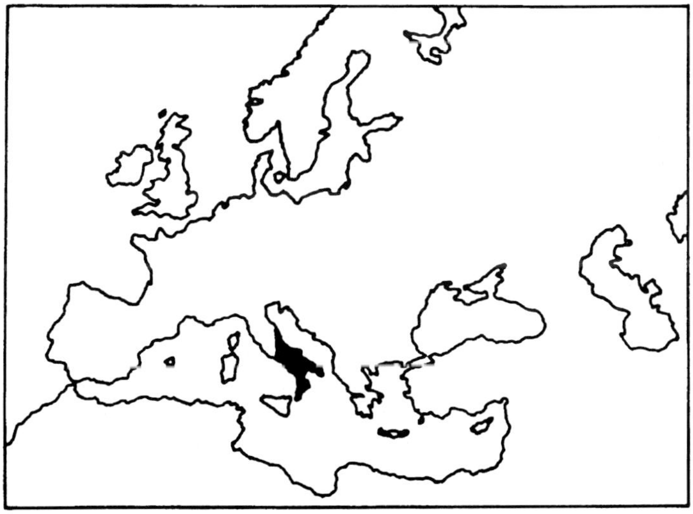

Distribution: Central and southern Italy.
Description: This is the smallest species of European newt, measuring up to 8 cm, males usually slightly smaller. In both sexes the body is greenish-brown or olive-brown on the upper surface. The male is marked with large dark spots on the body and tail, but lacks toe flaps and a crest along the back. Ridges are present along each side of the back giving it a

flattened appearance. The tail has a dorsal fin and a very short tail filament. There is no dark stripe through the eye, but a small yellow patch is present behind the eye. The female has fewer and smaller spots. Both sexes have a pale yellow belly which contrasts with a more deeply pigmented throat. Both belly and throat have a variable number of dark spots, sometimes forming two diffuse rows along each side.
Habitat: Occurs up to about 1200 m, and breeds in a variety of water bodies, including ponds, ditches, slow-moving streams and lakes. Seems to prefer warm habitats and is often associated with woodland or agricultural land.
Habits: Very similar to the Smooth Newt in general behaviour. Tends to prefer the shallow edges of ponds where it avoids predation and feeds on a range of planktonic prey and other invertebrates.
Eggs and larvae: Indistinguishable from those of the Smooth Newt.
References: Giacoma & Sparreboom (1987); Joly & Giacoma (1992).

Triturus montandoni (Boulenger, 1880)

Carpathian Newt (Plate 11)

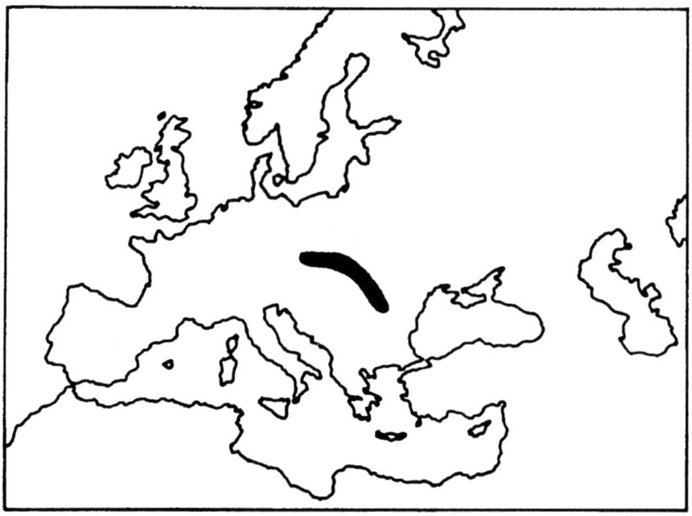

Distribution: Restricted distribution in southern Poland, Ukraine, Romania and Slovakia, where it occurs in the Carpathian, Tatra and Oder Mountains.
Description: Adults up to 10 cm long, males slightly smaller. Males are similar to male Palmate Newts in appearance; when in breeding condition they have a low, smooth crest, short tail filament, and ridges along the back which give a square shape to the body. The skin texture is slightly rougher than in other small-bodied newts, giving the surface a velvety texture even when in water. The body and tail are marked with scattered spots, which may extend to the edge of the belly. The lower edge of the tail has a whitish or bluish flash. The female is superficially like a female Smooth Newt. The colouration is olive-green or brown with a few fine dark spots on the back, sides and tail. Some females have darker markings which gives a marbled appearance. In both sexes the underside is a uniform

orange or orange–yellow, which grades to a paler shade on the throat. It is usually unspotted.
Habitat: Generally regarded as a montane species, breeding in small pools, ditches and slow-moving streams up to 2000 m altitude and often in association with deciduous woodland.
Habits: Typical pond-breeding newt, which enters water in March and returns to land in June. Where its distribution overlaps with the Smooth Newt hybridisation may take place.
Eggs and larvae: Eggs are similar to those of the Smooth Newt but may be more brownish. Larva is a typical pond-dwelling form, and very similar to *T. vulgaris.*
References: Pecio & Rafinski (1985).

Triturus vulgaris (Linnaeus, 1758)

Smooth or Common Newt (Plate 12)

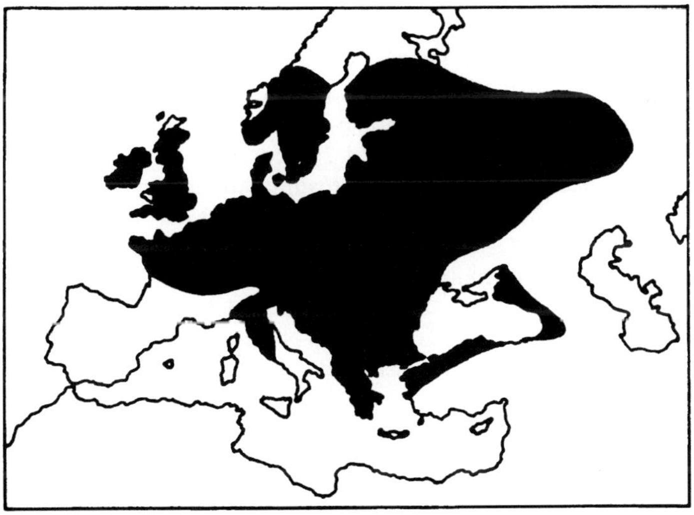

Distribution: Widely distributed over north-west, central and eastern Europe, including southern Scandinavia, western Asia, and Asia Minor. Absent from south-west France and Iberian peninsula. Only newt occurring in Ireland.
Subspecies:

Triturus vulgaris ampelensis: Northern Romania.
Triturus vulgaris graecus: Greece, Albania, Macedonia, Ionian Islands, Bulgaria.
Triturus vulgaris kosswigi: Turkey; south-west edge of the Black Sea.
Triturus vulgaris lantzi: West and central Caucasus along eastern edge of the Black Sea.
Triturus vulgaris meridionalis: North and central Italy, southern Switzerland, Slovenia.
Triturus vulgaris schmidtlerorum: Western tip of Turkey.
Triturus vulgaris vulgaris: Much of north-west, central and eastern Europe, western Asia and part of Asia Minor.

Description: Adults up to about 10 cm long, although the 'dwarf' subspecies (*T. v. schmidtlerorum*) may reach only half that size. Both sexes have a smooth, olivaceous skin.

Breeding males have a distinctive fleshy crest extending from behind the head on to the tail; flaps of skin on toes of hind feet; blue and red flashes on lower part of tail. At all times of the year males are marked with large black spots; belly has an orange or yellow flash with black spots; throat is white and distinctly spotted. Males of *T. v. graecus*, *T. v. kosswigi* and *T. v. meridionalis* have a much less wavy crest, a tail tapering to a fine point or filament, and wide toe flaps (they may therefore be confused with *T. helveticus*). Females are much drabber, and lack the crest, toe flaps and other breeding adornments of the male. Spotting on the belly and throat varies in intensity, but spots are always smaller than those of the male. Immature animals resemble females in general appearance, but may be distinguished from the closely related Palmate Newt by the dorsal stripe. In many animals this yellow stripe is most intense where it starts on the head, and then fades gradually along the back and does not reach the tail. Neotenous individuals of either sex are occasionally found. These may range from animals which have retained the full range of larval features but continued to grow without metamorphosing, to adults in full breeding condition which have just retained the larval gills.

Habitat: Occurs in a wide range of pond types in a variety of landscapes, where it is often the most abundant species of newt. Small to medium-sized pools which are fish-free may be preferred, although it may coexist with predatory fish in larger lakes by confining its distribution to the shallows. Terrestrial habitats include woodland, grassland, parkland, hedgerows, gardens, heathland and moorland. Indeed, this species does better than most in terms of establishing healthy populations in proximity to humans.

Habits: Spends most of the year on land, returning to water to breed in the spring or autumn (depending on subspecies and range). Males become sexually mature at 2–3 years old, females a year or so later. Adults live for up to 6–8 years. Males arrive at ponds before females, and stay in water for longer. Sex ratio of the population is variable, but rarely deviates from 1:1 by a large amount. In water Smooth Newts are most active during early morning and just after dusk, when most courtship takes place. On land they tend to stay within a general home area, where several hiding places may be regularly used.

Eggs and larvae: Between 200 and 300 eggs are laid individually, over a period of several weeks. The egg consists of a cream or amber-coloured embryo, surrounded by an ovoid jelly capsule about 4 mm in length. The larvae are typical pond-dwelling types, with stocky bodies and bushy gills. The tail has a pointed tip, and the toes are relatively short. Larvae avoid open areas of water and are usually found within vegetation or among detritus. They grow to just over 40 mm in length before metamorphosis. Those which hatch late in the year may continue to grow throughout the winter and metamorphose at a larger size the following spring.

References: Schmidtler & Schmidtler (1983); Griffiths (1984); Raxworthy (1990).

Triturus alpestris (Laurenti, 1768)

Alpine Newt (Plate 13)

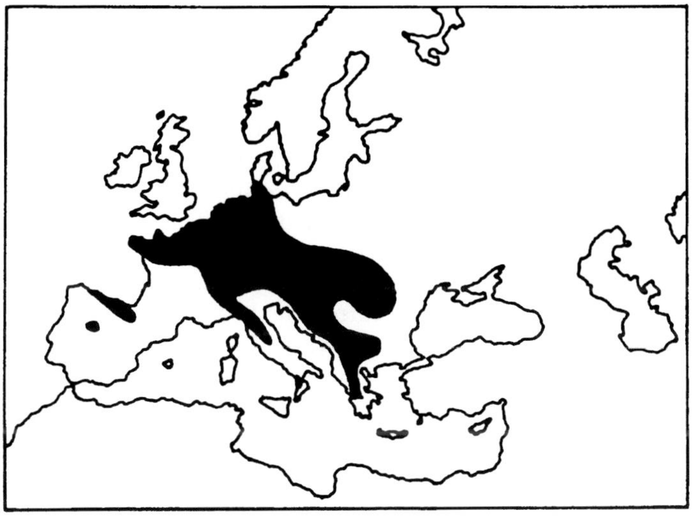

Distribution: Large area of central and southern Europe, from north-east France, through Poland and Romania to west Ukraine; and from southern Denmark to northern Italy, the Balkans and Greece. Two subspecies have isolated ranges: one in north and central Spain, the other in south and central Italy. Some introduced populations occur in Britain.

Subspecies:

Triturus alpestris alpestris: From central and eastern France eastwards to the Ukraine; and from southern Denmark southwards to the Balkan peninsula.

Triturus alpestris apuanus: South-east France and north-west Italy.

Triturus alpestris cyreni: Northern and central Spain (Sierra de Guadarrama, Segovia).

Triturus alpestris inexpectatus: Central Italy and Calabria, southern Italy.

Triturus alpestris lacusnigri: Julian Alps, Slovenia.

Triturus alpestris montenegrinus: Bukumirsko Lake, Montenegro (neotenous form).

Triturus alpestris piperianus: Montenegro (neotenous form).

Triturus alpestris reiseri: Vranika Mountains, Bosnia.

Triturus alpestris serdarus: Montenegro (neotenous form).

Triturus alpestris veluchiensis: Veluchi Mountains, Central Greece.

Description: A medium-sized newt measuring up to 10 cm; females slightly larger. Both sexes are distinctively marked and not easily confused with other species. Background colour is dark grey, blackish or bluish, with a bright orange or reddish belly and throat. Belly is unspotted, but there may be a few dark spots on the throat. Males have a whitish band with distinctive dark spots running from the cheeks, along the lower flanks, and on to the cloaca and tail. Between this and the orange belly, a light blue flash runs between the front and back legs. Breeding males have a low, smooth-edged crest which runs along the back and on to the tail. The crest is also whitish with regular dark spots which gives it a zig-zag or checkerboard effect along the back. Females are slightly drabber than males, and may have a green/blue mottled appearance. They also have the black-spotted and bluish flashes along the side, but these are usually less distinctive than in the male. The cheeks are also less spotted in the female, although there is a yellow stripe running along the lower edge of the tail which is absent in the male. The skin is slightly granular in texture. Juveniles are less brightly coloured than the adults but otherwise have the same basic colouration as females. Some of the subspecies are controversial. *T. a. inexpectatus*, for example, may be no different from *T. a. apuanus*.

Habitats: Although frequently associated with alpine and other montane habitats up to 2500 m, this species is also widespread, and often abundant, at lower altitudes. Occurs in a variety of ponds, lakes, ditches and slow-moving streams. These include both deep and shallow water bodies which are cool and shady, and in lowland areas it may be associated with woodland. In the Balkan States and Greece the habitats are often much more arid, with newts using small pools in dry, limestone areas.

Habits: In lowland areas this species may enter water in February (earlier in the southern part of its range), and stay aquatic until the autumn. Montane ponds may have ice cover for much of the year, and therefore a much shorter season for breeding. In such cases newts may enter water as late as June and leave in September. Adults tend to be bottom-dwellers, and are active mainly in the morning and evening.

Eggs and larvae: About 150 individual eggs are laid which hatch into pond-type larvae. These grow up to 60 mm before metamorphosis. The end of the tail usually has a very short filament. Neoteny occurs frequently in this species, with breeding newts retaining gills and other larval characteristics into adulthood.

References: Breuil & Guillaume (1985); Arano & Arntzen (1987); Schabetsberger & Jersabek (1995).

Triturus vittatus (Gray, 1835)

Banded Newt (Plate 14)

Distribution: Caucasus and Asia Minor from southern and eastern edges of the Black Sea in north Turkey southwards to Israel.

Subspecies:

Triturus vittatus ophryticus: The Caucasus; east and south of the Black Sea.

Triturus vittatus vittatus: Eastern edge of the Mediterranean Sea, including Israel, Lebanon, Syria and southern Turkey.

Description: A distinctively marked species which is not easily confused with other newts. Adults are 10–14 cm in length, and therefore a little smaller than the crested newts. Males in breeding condition are green, yellowish or light brown in colour, often with a metallic sheen, and with a fine, darker green or brown mottling over the head, body and legs. The species takes its name from a cream-coloured band which runs from the cheeks, along the lower flanks, and on to the tail. This band is attractively bordered by darker lines. The crest of the male is quite spectacular. It measures up to 45 mm high and has tooth-like indentations, and alternating light- and dark-coloured vertical bars. The tail is broad and less indented than the crest, and may have a bluish hue along the upper and lower edges. The cream-coloured band may persist along the middle of the tail but becomes broken up by a series of darker spots and rosettes on either side. The toes on the hind feet of the male have thin flaps of skin surrounding them. Females are much drabber in appearance, and slightly smaller than males. They are usually a uniform olive-green or brown, but retain the light-coloured band along the lower flanks. A fine white line, bordered by dark markings, runs from the base of the tail, but may fade rapidly. In both sexes the belly is yellow or orange, with darker markings towards the front of the body and on the throat. Sometimes spots also occur on the belly, but this is most frequent in *T. v. vittatus.*

Habitat: In the Caucasus this species is regarded as a montane amphibian, breeding in clear pools and streams with little vegetation but with an abundance of invertebrates. Further

south it also occurs at lower altitudes and uses a wide variety of water bodies, including temporary ponds (or even puddles), rock pools, streams and lakes. The terrestrial habitat may be quite arid, with vegetation consisting of Mediterranean oak, yew or beech.
Habits: In the north of its range this species breeds in spring and early summer. To the south the summers are very dry, and activity is related to the filling of temporary rainpools in the winter. In Israel adults may therefore leave the water as early as February or March, and the aquatic period is consequently shorter. The sex ratio may be close to unity or have a slight excess of males. Most courtship behaviour occurs after dusk. This newt is more territorial than any other European species. Males are intolerant of the presence of other males during the breeding period, and will chase and bite each other aggressively.
Eggs and larvae: Each female lays up to 150 individual eggs, over a period of up to 2 months. In warm, low-lying pools development and metamorphosis may be completed by May, while in cool, upland areas it will be much later in the year. The larvae are typical pond type, and tend to swim within the water column rather than on the pond bottom.
References: Degani & Mendelssohn (1983); Geffen *et al.* (1987); Raxworthy (1989).

Triturus carnifex (Laurenti, 1768)

Alpine Crested Newt (Plate 15)

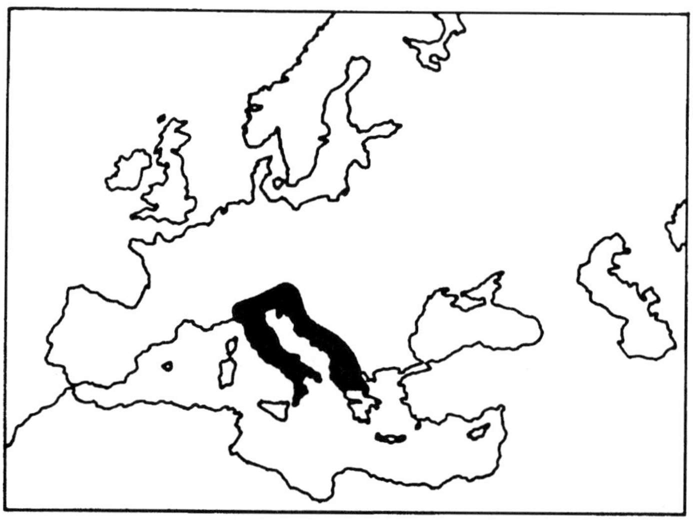

Distribution: Italy, southern Alps and the Balkans.
Description: Body of medium build, up to 17 cm, males 1–2 cm smaller. Upper surface is dark brown with black spots. Little or no white stippling on the flanks. Throat usually yellow and black with white stippling. Belly is orange with large, rounded blotches. Females and juveniles frequently have a conspicuous yellow stripe running along the back. Breeding males have a high indented crest running along the back and tail and a whitish flash on the tail. These features are not usually observable in individuals under 9.5 cm long. Wolterstorff Index (forelimb length/distance between limbs) is 0.53–0.64 for females.

Habitat: Occurs in a wide variety of both permanent and temporary water bodies. Sometimes occurs in quite arid areas of the Mediterranean, where water levels fluctuate and the terrestrial habitat may be harsh and dry.
Habits: Behaviour is much the same as *T. cristatus* (see below). Adults are found in most abundance in ponds from March until May when most breeding occurs, although low numbers of individuals may be found in water throughout the year in some places. Adults are mainly bottom-dwellers while larvae swim within the water column.
Eggs and larvae: Eggs are the same as those described for *T. cristatus* (see below), and suffer 50% mortality because of a chromosome anomaly during development. Larvae hatch around May and metamorphose at about 70 mm by September. Some populations contain neotenous newts.
References: Andreone & Giacoma (1989); Wallis & Arntzen (1989); Kalezic *et al.* (1994).

Triturus cristatus (Laurenti, 1768)

Northern Crested Newt (Plate 16)

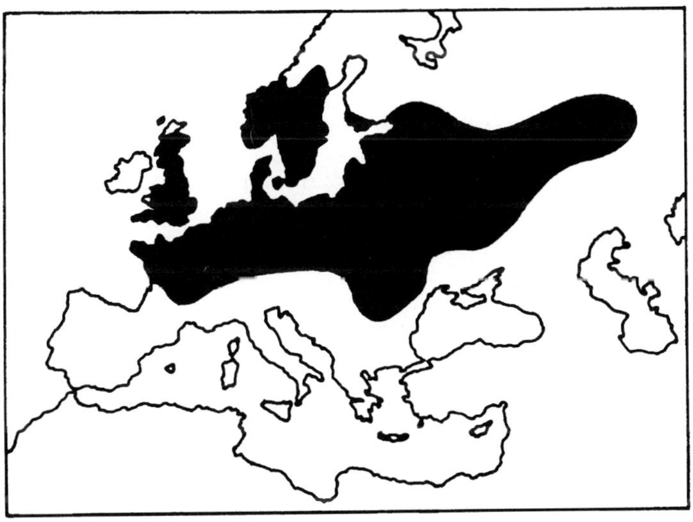

Distribution: Wide distribution from northern France, Britain and southern Scandinavia, through central Europe to the Ural Mountains in Russia.
Description: Moderately slender body, with a Wolterstorff Index (forelimb length/distance between limbs) of 0.45–0.6 for females. Up to 16 cm long, males 1–2 cm shorter. Dark brown, warty body, with black spots and conspicuous white stippling on flanks and legs. Throat is dark with white stippling. Belly is yellow/orange with irregular black blotches, which are most abundant towards the anterior end. Breeding males have an indented crest which extends on to the tail, and a whitish flash on the sides of the tail.
Habitat: Prefers medium- to large-sized ponds, particularly mature farm ponds and those associated with mineral workings. However, it will also use smaller water bodies, including ditches, shallow temporary ponds that dry out in the summer, and sometimes garden ponds. Optimal terrestrial habitat is characterised by plenty of ground cover, in the form of trees, scrub, long grass, stones and log piles.

Habits: Usually overwinters on land, and enters the breeding ponds in late February or early March. Males develop their breeding adornments very quickly and breeding can occur from March until June. Sex ratio is usually close to 1:1, and males adopt temporary territories in clear areas on the pond bottom where they display to females and other males. Adults leave the water in the autumn. Sometimes there is a migration towards the ponds in the autumn as well; this often comprises young animals which will breed for the first time in the following spring. On land newts may wander as far as 1 km from the breeding pond. Crested Newts are nocturnal amphibians and spend most of the time on the pond bottom. Immature individuals may return to water each year until adult, which is usually achieved at 2–3 years of age.
Eggs and larvae: About 200 eggs are laid individually, and consist of a pale yellow embryo about 2.5 mm long, surrounded by an ovoid jelly coat. They are carefully wrapped in the leaves of floating or submerged water plants. Half of the eggs fail to hatch because of an inherent fault during development. Eggs hatch into larvae which are about 12 mm long, and grow to a size of about 70 mm before metamorphosis. Unlike the adults, the larvae swim actively within the water column. Well-developed larvae have a conspicuous tail fin which tapers to a fine point, and elongate, slender toes. There are conspicuous dark spots on the body and tail. The head is rounded and rather flattened, and has three pairs of bushy gills behind it. Metamorphosis occurs towards the end of the summer or early autumn, although some larvae may overwinter in water, continue to grow, and metamorphose the following spring.
References: Griffiths & Mylotte (1987); Wallis & Arntzen (1989); Arntzen & Teunis (1993).

Triturus dobrogicus (Kiritzescu, 1903)

Danube Crested Newt (Plate 17)

Distribution: Two areas to the west of the Black Sea: (1) Romania, Moldavia and south-west Ukraine; (2) Hungary, east Austria, south Slovakia, Slovenia, Serbia and north Croatia.

Description: Up to 16 cm, males about 1.5 cm smaller. This is the most slender of the *cristatus* 'species group'. Consequently the body is long in relation to the length of the forelegs, which gives a Wolterstorff Index (forelimb length/distance between legs) of 0.34–0.45 for female newts. Dark brown body with black spots and white stippling on the flanks. Black throat with fine white spots; belly deep orange with angular black blotches. Breeding males have a crest which may extend to the front of the head. This is deeply indented along the back, less so on the tail.
Habitat: Typical habitats are flooded lowlands with lakes and oxbows, although it also occurs in ponds, ditches and sometimes rivers. May occur in disturbed habitats or in those close to urban development, and frequently co-exists with fish.
Habits: Migrations to breeding ponds start between mid-February and the end of March. Sometimes there is a second migration back to the pond in autumn. Both adults and subadults return to water and may remain aquatic for 6 months or longer. Adults may live for up to 8 years or more.
Eggs and larvae: Eggs and larvae are the same as those described for *T. cristatus* (see above), and suffer 50% mortality because of a chromosome anomaly during development. The first juveniles metamorphose and leave the pond between the end of August and the end of September.
Reference: Wallis & Arntzen (1989); Thonke *et al.* (1994).

Triturus karelinii (Strauch, 1870)

Southern Crested Newt (Plate 18)

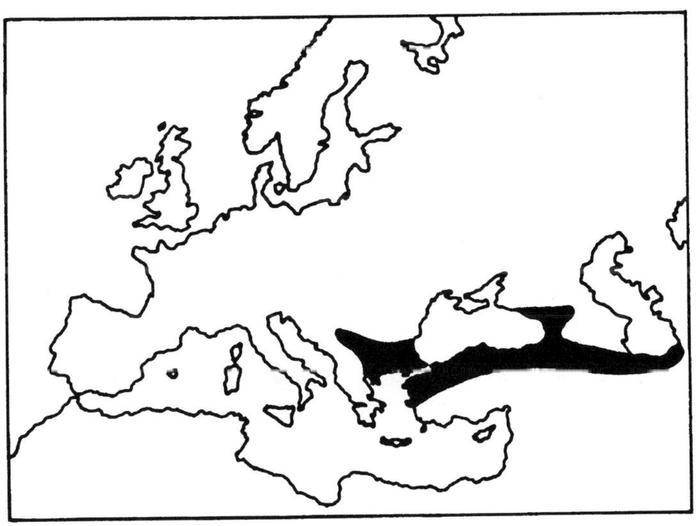

Distribution: Around the west, south and east edges of the Black Sea; extending from Serbia in the west to the Caspian Sea in the east, and south into central Turkey.
Description: The largest and most stockily built newt in the *cristatus* 'species group', with a Wolterstorff Index (forelimb length/distance between limbs) of 0.55–0.70 for females. Up to 17.5 cm, males slightly smaller. Body is dark brown with black spots. Some white

stippling on the flanks, with more conspicuous whitish markings on the cheeks. Belly is orange with many small or medium-sized black spots; unlike other crested newts, the belly colouration continues on to throat. Breeding males have a moderately indented crest which is smoother on the tail. Tail has a whitish flash.
Habitat: More of an upland species than the other crested newts, this species breeds in a variety of ponds, lakes and ditches.
Habits: Although this is a little-studied species, general ecology and behaviour seems to be very similar to that of the other crested newts.
Eggs and larvae: Eggs are the same as those described for *T. cristatus*, and suffer 50% mortality because of a chromosome anomaly during development.
Reference: Wallis & Arntzen (1989).

Triturus marmoratus (Latreille, 1800)

Marbled Newt (Plate 19)

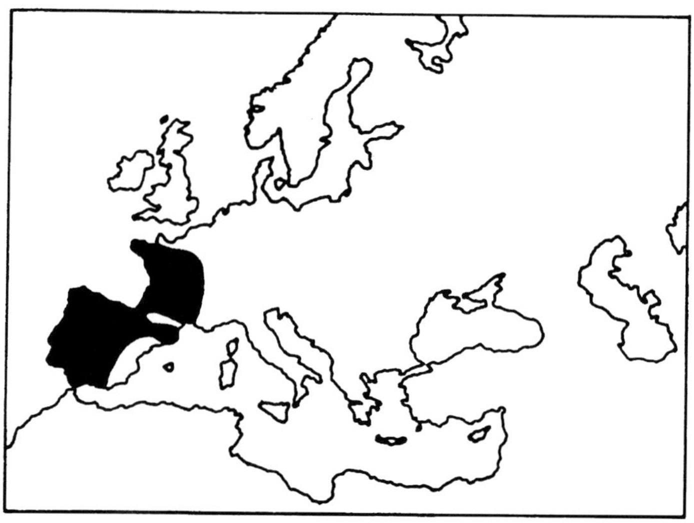

Distribution: South-west Europe: Iberian peninsula and most of France.
Subspecies:
Triturus marmoratus marmoratus: North Iberian peninsula, most of France.
Triturus marmoratus pygmaeus: South-west Iberian peninsula.
Description: A distinctive species which is difficult to confuse with any other newt. A large-bodied newt up to 16 cm; males 1 cm or so smaller than females. The skin is warty and has a bright green background colouration with irregular black blotches. Females may have an orange stripe running down the middle of the back and on to the tail. Breeding males develop a fleshy crest with distinctive vertical black bars. The crest extends on to the tail, which, like the crested newts, has a silvery stripe. In terrestrial males the crest regresses but a vestige is still visible as a black and yellow banded stripe running down the back. In both sexes there may be fine white stippling on the lower flanks. In *T. m. marmoratus* the underside is blackish with fine white spots. In *T. m. pygmaeus* there are also yellowish markings

and larger white spots on the same area. Juveniles are similar to females in colouration but often more brightly coloured. *T. m. pygmaeus* is about 3 cm smaller than *T. m. marmoratus* with males having a lower and straighter crest. Some authors consider *T. m. pygmaeus* to be a full species, *T. pygmaeus*.

Habitat: Usually found in well-vegetated ponds below 400 m altitude, but does occur in some undulating or hilly areas with plenty of cover. Although often associated with woodland or forests, the Marbled Newt breeds in a wide variety of pools, including cattle troughs, quarry ponds, ditches, rock pools in small streams, and ponds in fields or rough grassland.

Habits: In northern Europe the breeding period lasts from March until May, but may start much earlier in the southern part of the range. During courtship males seem to be more aggressive towards each other than Crested Newts and display vigorously in open areas on the pond bottom. The range of the Marbled Newt overlaps with that of the Crested Newt in central and north-west France, and in some ponds the two species hybridise. The hybrids, which were once considered to be a separate species *T. blasii*, may also breed with each other or the parent species, but the viability of their embryos is low.

Eggs and larvae: The Marbled Newt may lay about twice as many eggs as the Crested Newt (i.e. up to about 400 per year) but like the Crested Newt, half of these fail to develop beyond the tail-bud stage because of a genetic problem in development. Like Crested Newts, Marbled Newt larvae grow to a size of about 70 mm and have a tail which tapers gradually to a fine point. The tail fin has dark spots, but the body is usually less spotted than that of the Crested Newt.

References: Schoorl & Zuiderwijk (1981); Arntzen & Hedlund (1990); Garcia-Paris *et al.* (1993).

Euproctus asper (Dugès, 1852)

Pyrenean Brook Salamander (Plate 20)

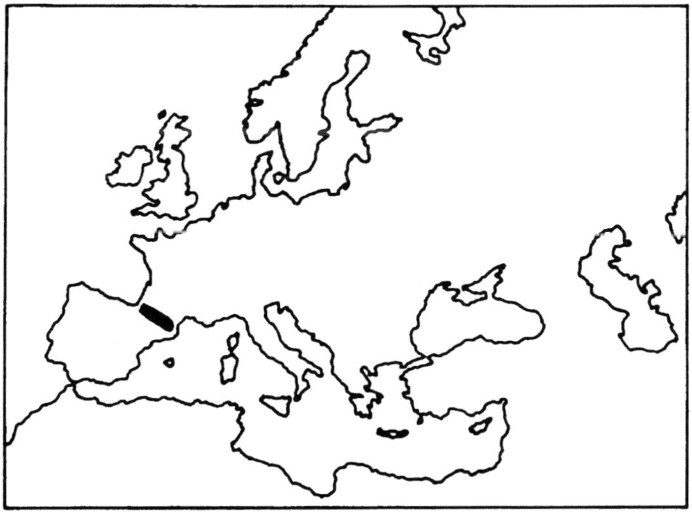

Distribution: Pyrenean Mountains, particularly the east and central sections.

Subspecies:

Euproctus asper asper: Occurs over most of the range for the species.

Euproctus asper castelmouliensis: Confined to the Torrent de Castelmouly, in the French Pyrenees.

Description: This is the largest and most stoutly built of the brook salamanders, measuring 10–14 cm long, and is also distinguished from other species by its rough skin. Background colour is dark grey, brown or black, often with a few scattered yellow spots. There may also be a yellow stripe running down the back and on to the upper edge of the tail, but this may be broken into a series of irregular blotches. However, colouration seems to vary considerably between populations. The tail is flattened from side to side. The belly is yellow or orange and may be unspotted. Unlike other salamanders there are no parotoid glands behind the eyes. Females are similar to males in general appearance, but can be distinguished by their pointed, rather than rounded, cloacal opening. The status of *E. a. castelmouliensis* as a separate subspecies is controversial.

Habitat: A high-altitude salamander, usually found at heights of over 600 m. Occurs in cold mountain lakes, streams, and occasionally cave systems, which are ice-free for at least 4 months of the year.

Habits: Adult salamanders are most active in water between the spring and the autumn. During courtship the male signals to the female with his tail raised in a near-vertical position. He then restrains her by wrapping his tail around her body. This position may be maintained for several hours while spermatophores are transferred to the female's cloaca.

Eggs and larvae: Clutch sizes are small, with each female laying between 20 and 40 individual eggs between April and July. These are deposited under rocks or between stones, and hatch into larvae which are about 13 mm long. The larva is a typical stream type, with a streamlined body, small gills, and low tail fin with a rounded tip. It metamorphoses when it reaches a size of 50–60 mm.

References: Garcia-Paris (1985); Thiesmeier & Hornberg (1990).

Euproctus montanus (Savi, 1838)

Corsican Brook Salamander (Plate 21)

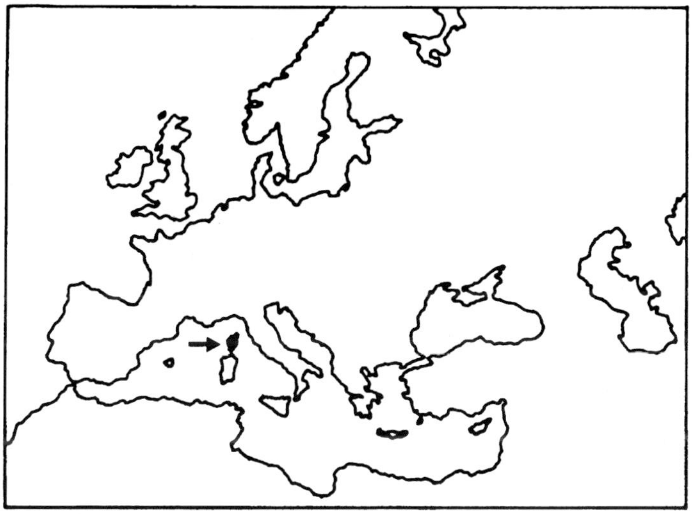

Distribution: Confined to the Island of Corsica.

Description: Smaller than the Pyrenean Brook Salamander, this species measures up to 9–11 cm. It is smooth-skinned, and with distinct parotoid glands behind the eyes which gives the head a broad appearance towards the neck region. Background colour is olive green or brown, sometimes mottled with lighter or darker markings. A thin stripe running down the back may be present. The underside is grey or brown in colour and may have darker spots although the throat is usually unspotted. The tail is slightly flattened from side to side, although less so than in the Pyrenean Brook Salamander. Males can be distinguished from females by the presence of small spurs on the hind legs and a conical cloacal swelling.

Habitat: Occurs in rocky streams and brooks at altitudes above 600 m, although it can occasionally be found at sea level. Also occasionally occurs in spring-fed cattle troughs.

Habits: This species is active in water mainly during the spring and autumn months. During copulation the male wraps his tail around the female and grips her with his legs and mouth. Spermatophores are then transferred directly from the male cloaca into the female cloaca.

Eggs and larvae: Up to 70 eggs are laid individually, but in groups, under rocks and stones. Females stay close to the eggs during their development. Larvae are of the stream type with streamlined bodies, small gills and a rounded tail tip. They grow to about 50 mm before metamorphosis. At low altitudes larvae will metamorphose after a few months in the same year that they were deposited. At higher levels larval development may take longer than a year.

References: Alcher (1981a, 1988).

Euproctus platycephalus (Gravenhorst, 1829)

Sardinian Brook Salamander (Plate 22)

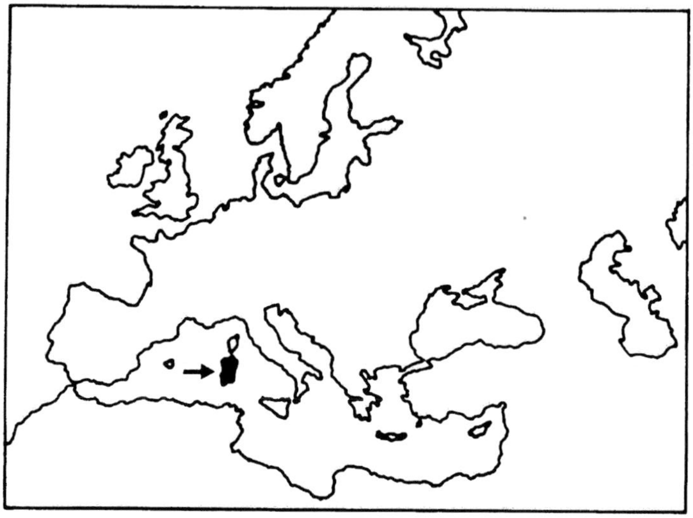

Distribution: Confined to the Island of Sardinia.
Description: Measures 10–14 cm, males slightly larger than females. Background colour is a light brown or olive-brown, with darker grey or green markings. The belly and throat are white, yellowish or grey in colour with black spots which are usually more conspicuous in males. The head is flatter and broader than the other brook salamanders, and although parotoid glands are present, they are not conspicuous. Both sexes have conically shaped cloacal swellings, but can be distinguished by the presence of spurs on the hind feet of the males.
Habitat: Most frequently found in montane streams and lakes between 1000 m and 1800 m. Sometimes also found at lower altitudes in streams running through wooded slopes.
Habits: Most active during the spring, early summer and autumn. Mating behaviour is similar to that of the Corsican Brook Salamander, with the male gripping the female with his tail, legs and mouth. Spermatophores are then manipulated into the female cloaca with the hind legs. There seem to be equal numbers of males and females in populations, and one 40 m wide lake supported a population of between 180 and 445 individuals during the summer.
Eggs and larvae: The number of eggs in a clutch varies considerably, but is higher than in the other two brook salamanders, with an average of about 140 eggs per female. Eggs are deposited between stones or buried in sand. Newly hatched larvae are about 13 mm and grow up to 60 mm before metamorphosis. They are stream-type larvae with long bodies, short gills and a rounded tip to the tail. Metamorphosis usually takes place in the year following egg-laying.
References: Alcher (1980, 1981b); Voesenek *et al.* (1987).

Pleurodeles waltl Michahelles, 1830

Spanish Ribbed Salamander (Plate 23)

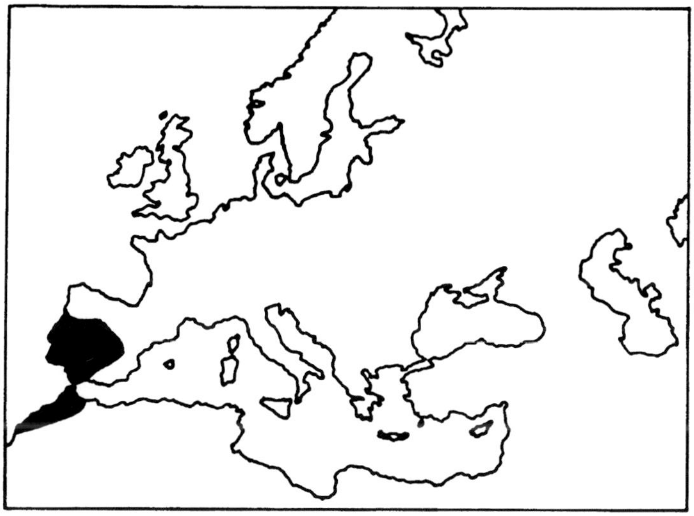

Distribution: Central and southern Iberia, Morocco. A closely related, but poorly studied species, *Pleurodeles poireti*, occurs in Tunisia and Algeria.
Description: A very large and stockily built salamander, up to 30 cm long. The skin is rough and olive-green, brown or yellowish in colour, with a lighter yellow, grey or whitish underside. Upper and lower surfaces are marked with darker blotches. Along each upper side of the body are two rows of yellowish warts, each of which covers a protruding rib. The pointed ribs afford the salamander some degree of protection against predators. The head is broad and flattened and the eyes relatively small. The limbs are short and stocky, and the tail is compressed from side to side. Males tend to have slightly longer and deeper tails than females, and rough nuptial pads on the front legs. They also have a reddish hue to the body.
Habitat: Ponds, lakes, ditches and slow-moving streams, usually with plenty of vegetation cover.
Habits: This is a very aquatic salamander (also known as the Sharp-ribbed Newt), that is rarely seen moving about on land. If the water body dries up it tends to bury into the bottom mud or crevices until the rains refill the pond. Breeding usually starts in March and is finished by June, but if winters are warm and wet it may begin as early as January, and sometimes continues for most of the year. When in water this species is a nocturnal bottom-dweller, and rather less active than other newts or salamanders. However, it is a voracious feeder and will consume almost any moving prey that comes within range and is small enough for it to swallow, including other amphibians. In courtship the male crawls under the female and locks his front legs over hers. After several hours in this position, he twists round and deposits a spermatophore, and then manipulates the female's body so that this is taken into her cloaca.
Eggs and larvae: About 200–300 eggs are laid individually, or in small clumps of up to 20 eggs, and attached to plants or stones. These hatch into typical pond-type larvae, which

grow to a size of around 10 cm before metamorphosis. When raised at high temperatures the sexes of the larvae are reversed. The larvae are stockily built with a high tail fin which tapers to a blunt point. The body and tail may be marked with dark spots.
References: Garcia-Paris (1985); Diaz-Paniagua (1992); Dorazi *et al.* (1995).

Salamandrina terdigitata (Lacépède, 1788)

Spectacled Salamander (Plate 24)

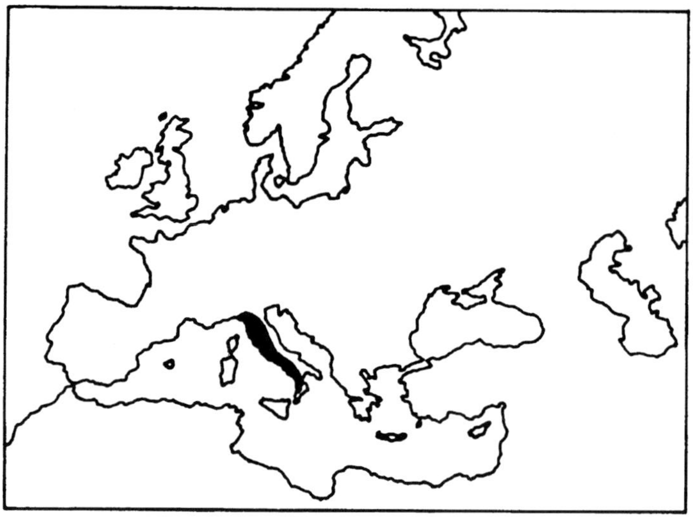

Distribution: Confined to the Appennines in west and central Italy.
Description: This species is unlikely to be confused with any other newt or salamander. Small, slender salamander (up to 10 cm), with a cylindrical tail which is longer than the body. Upper surface is a uniform dark brown or black with a conspicuous light-coloured 'V'-shaped patch on the head between the eyes. Skin is slightly rough to the touch with prominent ribs. The belly and throat are whitish with conspicuous black markings, which may almost cover the throat. The underside of the limbs and tail are strikingly coloured with red, which sometimes extends on to the belly. Males are slightly smaller than females but are otherwise similar in appearance. The only European salamander with four toes on the hind feet.
Habitat: Lives amongst leaf litter on wooded mountain slopes close to streams and seepages.
Habits: This is a secretive, little-studied species with poorly understood behaviour patterns. It is probably only active after dark and forages amongst the leaf litter or in other well-shaded areas in damp weather. It is a terrestrial species with courtship taking place on land, although females enter water to lay eggs.
Eggs and larva: Eggs are deposited in small groups in slow-moving streams or other clear pools of water. The larva is a typical pond type, and easily confused with young Fire Salamander larvae. However, well-developed Spectacled Salamander larvae are more slender and smaller (up to 30 mm) than those of the Fire Salamander.

Salamandra atra Laurenti, 1768

Alpine Salamander (Plate 25)

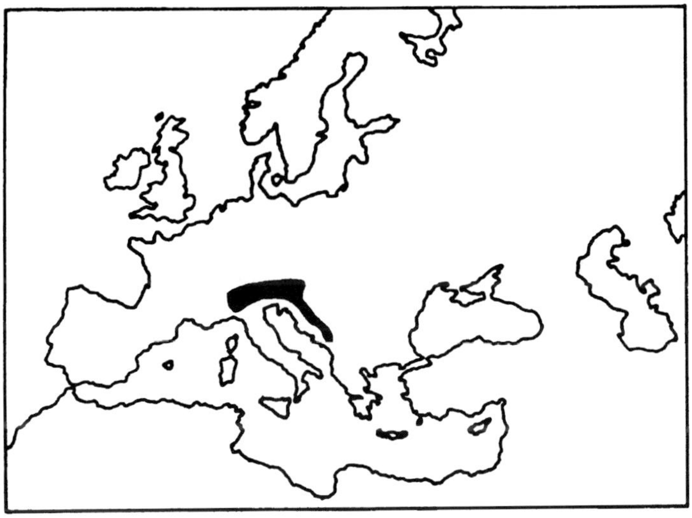

Distribution: Alpine regions in central Europe, ranging from the Franco-Swiss border, through Switzerland and Austria, north-east Italy, and into the western Balkans.

Subspecies:

Salamandra atra atra: Most of the area described above.

Salamandra atra aurorae: known only from altitudes of 1300–1500 m in the Bosco del Dosso in north-east Italy (province of Vicenza).

Description: Up to 14 cm long, and more slender than the Fire Salamander. Jet black, apart from *S. s. aurorae* which has straw-yellow markings on the head, extending in a broad band down the middle of the back. Further spots may occur on the tail and limbs. Two large parotoid glands behind the eye and a double row of poison glands running down the centre of the back and on to the tail. A further single row of poison glands runs along either side of the body. Males are similar to females but have a more prominent cloaca.

Habitat: Found between 450 and 3000 m in cool, damp alpine locations. Occurs in mixed woodland at lower altitudes but is also found in more open habitats above the tree-line providing there are stones, logs and a rich undergrowth of grasses and mosses to provide hiding places.

Habits: Tends to replace the Fire Salamander at high altitudes. Spends most of its time in cracks, crevices or burrows, and only emerges from its hiding places at night and after rain, when it can be surprisingly abundant. Alpine Salamanders are viviparous and entirely terrestrial, giving birth to just two fully metamorphosed young.

References: Trevisan (1982); Klewen (1991); Grossenbacher (1994).

Salamandra lanzai Nascetti, Andreone, Capula & Bullini, 1988

Lanza's Alpine Salamander (Plate 26)

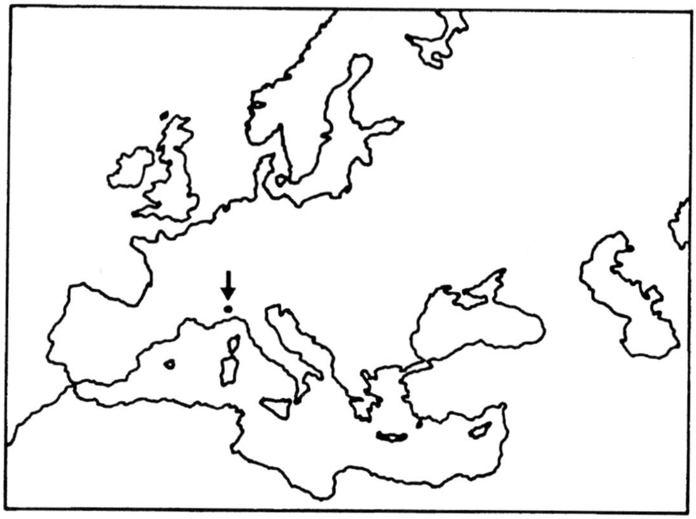

Distribution: A restricted area, Cottian Alps in west Piedmont between 1500 and 2000 m, south-west of Turin in north-west Italy. Possibly also the Maritime Alps.
Description: Very similar to *Salamandra atra*, but is slightly larger (up to 16 cm) and has a flatter head. Uniform black on upper surface, dark grey underneath. Poison glands are concentrated in parotoid glands behind each eye, and single rows on either flank, but there are none running down the back (cf. *Salamandra salamandra* and *S. atra*).
Habitat and habits: As *Salamandra atra*.
References: Nascetti *et al.* (1988); Andreone (1992); Grossenbacher (1994).

Salamandra salamandra (Linnaeus, 1758)

Fire or Spotted Salamander (Plate 27)

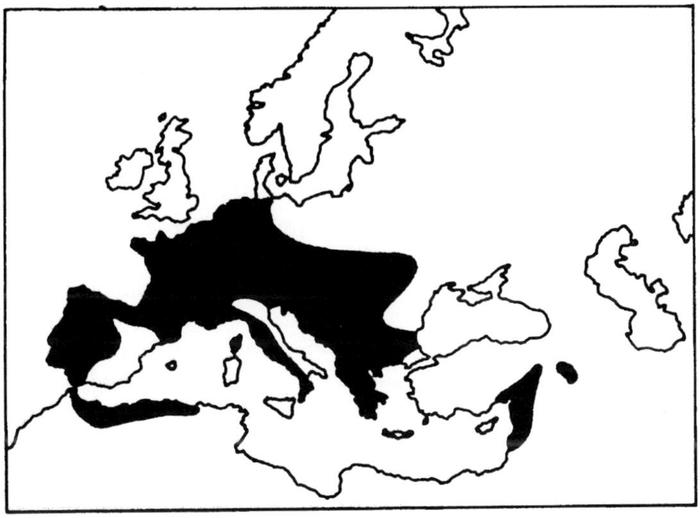

Distribution: Widespread across central and southern Europe, extending into north Africa, Turkey and the Middle East. Absent from Britain and Ireland, and Scandinavia.

Subspecies:

Salamandra salamandra algira: North Morocco and Algeria.

Salamandra salamandra almanzoris: Central Spain; Sierra de Gredos.

Salamandra salamandra bejarae: Iberian peninsula, except for north, east and west fringes.

Salamandra salamandra bernardezi: Northern Spain near Oviedo.

Salamandra salamandra beschkovi: Pirin Mountains, southern Bulgaria.

Salamandra salamandra corsica: Corsica.

Salamandra salamandra crespoi: Serra de Monchique, southern Portugal.

Salamandra salamandra fastuosa: Northern Spain including Pyrenees and Cantabrian Mountains.

Salamandra salamandra gallaica: Portugal and north-west Spain.

Salamandra salamandra giglioli: Central and west Italy.

Salamandra salamandra infraimmaculata: Western edge of the Mediterranean, including south-east Turkey, Syria, Lebanon and Israel.

Salamandra salamandra salamandra: From south-east France through central Europe and the Balkan peninsula to the Black Sea.

Salamandra salamandra semenovi: Zagros Mountains; eastern Turkey.

Salamandra salamandra terrestris: Central Europe, from north-east Spain, through France to central Germany.

Description: Stockily built salamander from 15 to 25 cm long, with specimens at the south-east corner of its range reaching nearly 30 cm. Background colour black, with yellow or orange spots forming distinctive patterns, usually extending on to the head, tail and limbs.

Yellow pigmentation may form a wide variety of patterns (depending on subspecies) ranging from large numbers of widely scattered spots to two, broad yellow bands running down either side of the body (Fig. 8.1). However, because of wide variation in the colour pattern, this feature is not an entirely reliable guide to subspecies identification. Some specimens may be almost entirely yellow; occasionally specimens with red pigmentation are found (especially in *S. s. gallaica*). Underside is dark grey and usually more lightly spotted, but some forms have extensive yellow markings extending on to the throat, belly and tail. Tail is cylindrical and slightly shorter than the body; limbs are stout with short, stubby toes. Conspicuous parotoid glands behind the eyes are always pigmented, and two rows of poison glands run down the centre of the back and on to the tail. Two more rows of poison glands run along each flank. Females are slightly larger than males, and males have a more swollen cloaca, particularly during the breeding period. Otherwise the sexes are very similar. Recent genetic analyses suggest that the subspecies *algira, corsica* and *infraimmaculata* may warrant full species status.

Habitat: Occurs in a variety of habitat types, but is always associated with shade and moisture. Preferred habitats are sheltered hillsides with deciduous or mixed woodland. In such areas it occurs under logs, stones, in holes around the bases of trees, or in other cracks and crevices that provide shade and mositure. Usually found in the vicinity of ponds or streams as these are required for breeding.

Habits: Although widespread and quite abundant in places, the Fire Salamander is secretive in nature and spends much of its time hiding away in refuges. It is only active on the surface on mild, damp nights. Extremes of temperature are therefore avoided, and in northern Europe salamanders are inactive during the winter, while in Israel activity ceases during the hot summer period. Similarly, in northern and central Europe breeding occurs between the spring and autumn, while at the southern edge of its range this activity is confined to the winter. Adult Fire Salamanders are almost entirely terrestrial. Mating occurs on land and females only immerse their hindquarters in a pond or stream to release the larvae.

Eggs and larvae: Fire Salamanders give birth to well-developed larvae. An exception is *S. s. bernardezi* which gives birth to fully metamorphosed salamanders (*S. s. fastuosa* sometimes shows the same pattern). Clutch sizes vary in size, although something between 10 and 30 is average. Much larger clutches are laid towards the south-east edge of the species' range. The larvae are typical pond type, and quite voracious, readily consuming their smaller sized counterparts. They grow to about 75 mm before metamorphosis, and have a tail which tapers to a rather blunt tip. The yellow and black colouration is not fully developed until after metamorphosis, although the markings start to be discernible in well-grown larvae.

References: Klewen (1991); Thiesmeier (1992); Veith (1994).

A

B

C

D

E

F

FIG 8.1 *Variation in the spot pattern of the Fire Salamander. A,* S. s. salamandra; *B,* S. s. bejare; *C,* S. s. bernardezi; *D,* S. s. fastuosa; *E,* S. s. gallaica; *F,* S. s. gigliolii. *(M. Garcia-Paris, R. Griffiths, B. Thiesmeier, P. Wisniewski).*

Mertensiella caucasica (Waga, 1876)

Caucasian Salamander (Plate 28)

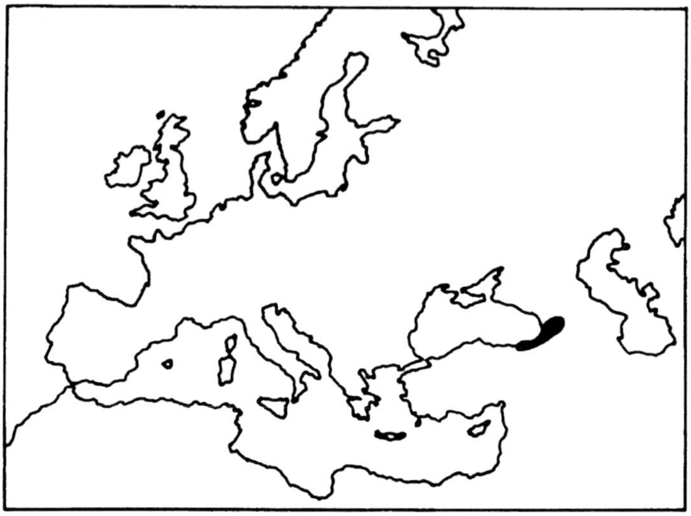

Distribution: North-east Turkey bordering the south-east edge of the Black Sea, and west Georgia.

Description: Up to 18 cm, this is a very slender salamander with a tail that is considerably longer than the body in adults. Background colour varies from reddish-brown to dark brown or black with a variable number of small yellow spots on the head, back, limbs and tail. The spots are sometimes arranged in two distinct lines running down the back; other individuals are almost entirely black. The underside is dark grey with fine, lighter speckling. Like Luschan's Salamander, males have a fleshy protuberance at the base of the tail although this is less pointed in appearance. Males may also have swollen forearms, which like the protuberance, are used in mating.

Habitat: A montane salamander found in the vicinity of streams running through humid forests and subalpine meadows. Here it lives in cracks and crevices underground, in log or stone piles, or under tree butresses. Also inhabits caves with underground streams.

Habits: The Caucasian Salamander spends most of its life in underground shelters close to streams, and only occasionally ventures on to the surface on cool, damp nights. Most activity occurs during the breeding period from May to July. Courtship behaviour consists of the male pursuing the female, and locking his front legs over hers after crawling underneath her body. Courtship may start in water but finishes with spermatophore transfer on land. Despite its secretive life-style, the Caucasian Salamander can move remarkably quickly on land even at low temperatures, and may drop its tail in a lizard-like fashion if attacked by a predator. This species appears to be long-lived: newly metamorphosed salamanders take around 10 years to reach sexual maturity and live for 25 years or more.

Eggs and larvae: During the summer females lay around 10–20 individual eggs, each about 5–6 mm in diameter. These are positioned so that they adhere to the substrate in sheltered places along the stream. Eggs have a long development time and can take 60 days or more

to hatch. Larvae hatch at about 15 mm, and have streamlined bodies typical of stream-dwelling salamanders. There are small, non-pigmented patches along the back and tail. Larvae spend most of the time hidden under stones or other cover objects, and do not complete metamorphosis until their third year of development when they have reached a size of about 35 mm.
References: Tarkhnishvili & Serbinova (1993); Schultschik (1994); Tarkhnishvili & Gokhelashvili (1994).

Mertensiella luschani (Steindachner, 1891)

Luschan's Salamander (Plate 29)

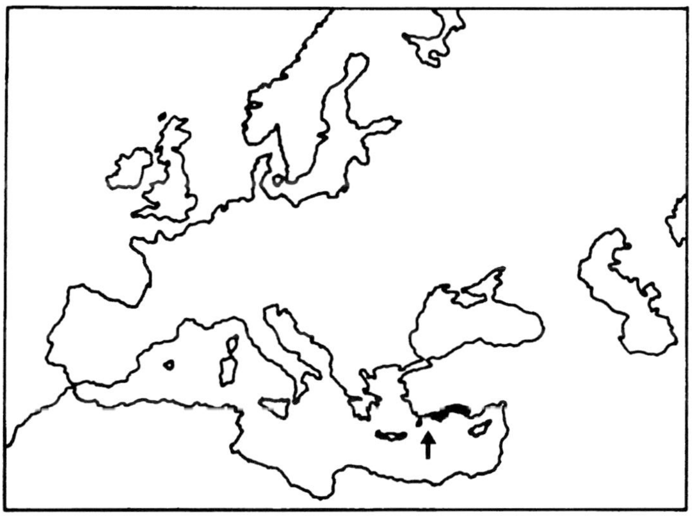

Distribution: South-west Turkey and the south-east Aegean islands of Karpathos, Saria, Kasos and Kastellorizo. The subspecies have small ranges, and some are isolated from each other over short distances by physical barriers.
Subspecies:

Mertensiella luschani atifi: Near Alanya, south Anatolia.
Mertensiella luschani antalyana: Antalya, south-west Anatolia.
Mertensiella luschani basoglui: Kas, south-west Anatolia.
Mertensiella luschani billae: South-west of Antalya, south-west Anatolia.
Mertensiella luschani fazilae: North-west of Fethiye, south-west Anatolia.
Mertensiella luschani finikensis: Finike, south-west Anatolia.
Mertensiella luschani helverseni: Islands of Karpathos, Saria and Kasos.
Mertensiella luschani luschani: South-east of Fethiye, south-west Turkey.

Description: This salamander measures up to 15 cm, with females being slightly larger than males on account of their longer tails. The different subspecies are similar in size, although *M. l. atifi* is the largest and *M. l. billae* the smallest. Compared to the Caucasian Salamander, the tail is shorter (being rather less than the trunk length), the head is broader and the toes stubbier. In males there is a pointed, fleshy protuberance at the base of the tail, which is oriented towards the head. Males may also have slight swellings on the forelimbs during the breeding period but these are not as conspicuous as in the Caucasian Salamander. There are pronounced but narrow parotoid glands behind the eyes which have orange pigmentation. Colouration varies considerably, and the upper surface may be orange, brown, reddish, grey or greenish with darker blotches and mottling. The darker markings may form two distinct bands running along the back. These markings grade into much lighter colouration along the flanks and underside. Lighter, yellowish spots are also usually scattered over the upper surface. The markings vary in distribution and intensity between subspecies. The skin is not entirely smooth as tiny, spiny projections can be felt along the back and on the throat.

Habitat: Occurs from sea level up to 1000 m in Mediterranean maquis, pine woodland and scrub. Shady slopes or valleys are preferred, particularly those with loose substrates and rocky crevices which provide an abundance of hiding places. It is also found in the vicinity of streams, and sometimes in surprisingly dry surroundings.

Habits: Luschan's Salamander can climb well, and during the hot summer months retires to hiding places in rocky screes or other cracks and crevices. Surprisingly fast moving for a salamander, mating probably takes place on land during the damp spring months. The male holds the female from below by locking his front legs over hers, and lowers her onto a spermatophore deposited on the substrate. The species is viviparous and gives birth to fully developed salamanders 30–50 mm long, between September and March, after a gestation period of at least a year. Sexual maturity is reached at a size of about 10 cm.

References: Klewen (1991); Baran & Üçüncü (1994); Polymeni (1994).

Chioglossa lusitanica Bocage, 1864

Golden-striped Salamander (Plate 30)

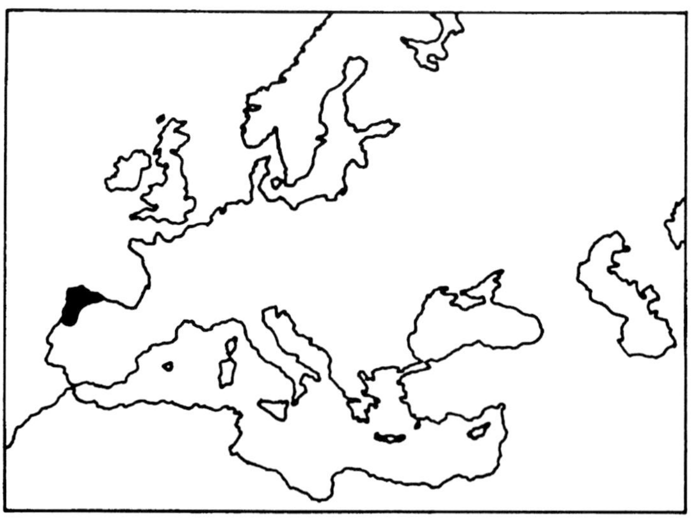

Distribution: North-west Spain and north and central Portugal.

Description: Total length up to 15 cm, including a long tail nearly twice the length of the body (though some individuals may have lost and then regrown much shorter tails as a result of predation). Very slender body, usually brownish in colour with a brown or grey underside. Typical specimens have two copper or golden stripes running down the back and joining at the tail. In some populations these stripes may be broken up by darker patches, or even reduced to a few golden spots. Males are distinguished from females by their swollen cloaca and rough pads on the upper arms. These features are easier to see in animals in breeding condition.

Habitat: Along the edges of rocky streams and brooks in hilly areas up to 1500 m altitude, and with over 100 cm rain per year. In damp areas it may live further away from flowing water.

Habits: This salamander leads a secretive life, and its annual cycle is dependent on climate. However, it can move remarkably quickly on land. Population densities vary within its range, but in some places it is the most abundant brook-dwelling amphibian. Adults are only active on land at night and after rainfall. The long tail may be used to provide purchase when climbing rocks on land or for propulsion when in water. Juveniles may be more day-active providing the ground is damp. If streams dry up then the salamanders will seek refuge in cracks and crevices underground. If attacked by a predator or roughly handled the tail may be shed. Mating can occur at almost any time of the year except for winter, but is most intense during summer and autumn. The sex ratio is close to 1:1.

Eggs and larvae: Eggs are spherical, about 5 mm in diameter, with a yellowish embryo and jelly coat. They are laid in small clusters in cavities or under stones in the stream. Egg deposition occurs mainly in the autumn, and the larvae hatch 6–9 weeks later. Most larvae therefore overwinter and metamorphose the following spring or summer; some larvae grow through two winters. Larvae are of the stream type, with streamlined bodies and short external gills. The end of the tail is rounded. They reach a size of about 50 mm before metamorphosis. Larvae live a sheltered life amongst vegetation or stones within the stream.

References: Busack (1976); Arntzen (1981); Vences (1990).

FAMILY: PROTEIDAE

This is another example of a family which has a very restricted distribution in Europe. Indeed, the family is represented by just two genera world-wide, one in Europe (*Proteus*) and one in North America (*Necturus*). The family consists of salamanders which are entirely neotenous, but which are otherwise rather poorly studied.

Proteus anguinus Laurenti, 1768

Blind Cave Salamander or Olm (Plate 31)

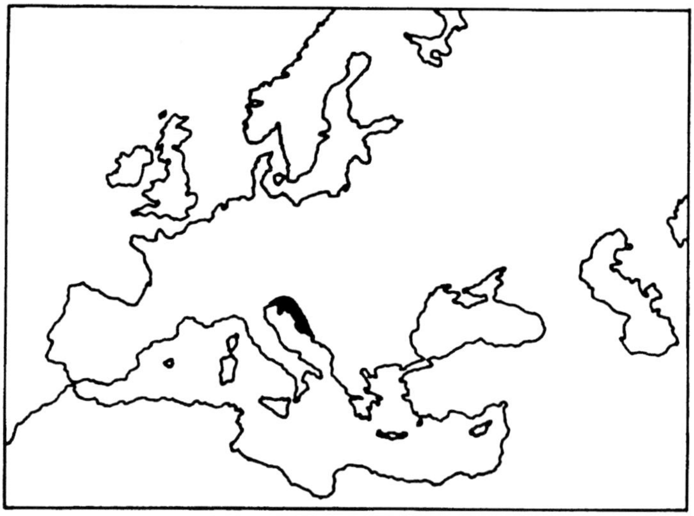

Distribution: The Dinaric karst region from north-east Italy, through Slovenia to the River Trebisnjica in Hercegovina.
Subspecies:
Proteus anguinus anguinus: Most of the above region.
Proteus anguinus parkelj: Bela Krajina, south-east Slovenia.
Description: This is a permanently aquatic, neotenous salamander which measures up to 28 cm long. Both subspecies have elongate bodies, weak limbs and a flattened tail which is rather shorter than the body. The skin is smooth with over 25 costal grooves along each flank. Three pairs of fluffy, red gills are positioned at the back of the head. The forelegs have three digits each while the hindlegs have only two. *P. a. anguinus* lacks pigment and apart from the gills is a uniform white or pale pink. The tiny eyes are completely covered by skin, but may be sensitive to light and darkness. The subspecies *P. a. parkelj* is black or dark brown with externally visible eyes and a shorter head. The limbs and tail are also shorter. The taxonomic status of the species is currently under review.
Habitat: Springs and underground streams in limestone regions.
Habits: Little is known about the habits of this poorly understood amphibian in the wild. Eggs, larvae and gravid females are rarely found in nature, and most observations stem from studies of captive animals. Courtship behaviour is rather like that of the European newts

with males displaying to females with tail-fanning movements. Competing males may chase or even bite each other. The male deposits a spermatophore which is taken up by the female so that eggs are fertilised internally. The white form spends its entire life in underground streams, and finds prey using water currents and chemical cues. The black form occasionally forages in above-ground streams, and may be able to use its eyes in foraging. Adults take at least 14 years to reach sexual maturity and are probably very long-lived.
Eggs and larvae: Captive animals lay up to 70 eggs, which are stuck to the underside of rocks and take several months to hatch. During this period the adult salamander stays near the eggs. The hatchlings are about 22 mm long and have prominent eyes which later degenerate. The front legs have a full complement of three digits while the back legs have none. It is also possible that under certain conditions this species is viviparous and produces just two well-developed young.
References: Briegleb (1962); Aljancic *et al.* (1993); Sket & Arntzen (1994).

FAMILY: HYNOBIIDAE

The inclusion of this rather ancient family of Asiatic salamanders in this book is warranted by one species which has a range which extends into Europe. The family consists of a group of small-bodied, slender salamanders which return to water to breed. Larvae hatch from eggs and pass through a typical aquatic development phase before metamorphosis. Lungs are small or absent, and some species are therefore confined to well-oxygenated streams or brooks.

Salamandrella keyserlingii Dybowski, 1870

Siberian Salamander (Plate 32)

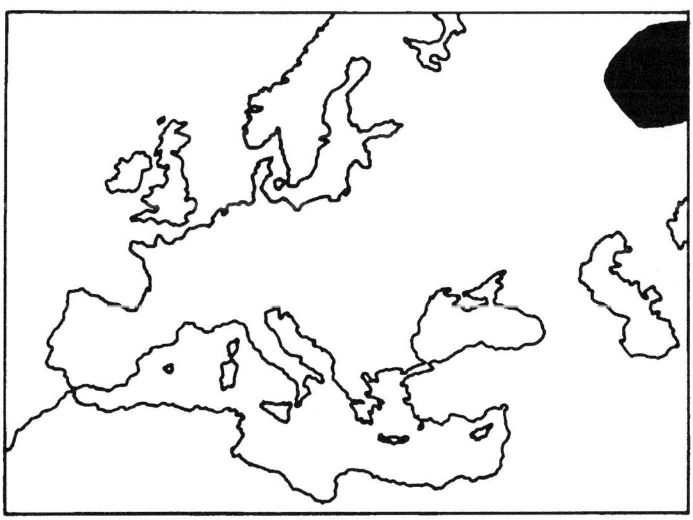

Distribution: A very widely distributed species which just enters European Russia. Its range extends from west of the Urals eastwards across Siberia to Kamchatka; and from the Yakutian tundras southwards to Mongolia, Manchuria and North Korea.

Description: This is a smooth-skinned salamander which measures up to about 13 cm in length, including a tail which is just shorter than the head and trunk. The head has a rounded snout, large eyes and prominent glandular cheeks. The body has a rather square appearance with about 14 vertical grooves (costal grooves) which continue on to the tail. The tail is paddle-shaped and flattened from side to side. Front and back limbs have four digits each. The top of the head and back are light brown in colour with a rather metallic sheen. There may be a thin, dark line running down the middle of the back. The edges of the back and the sides of the tail are mottled with darker markings, while the lower flanks, cheeks, and tail edge are greyish with fine dark spots. Females are slightly larger than males, but otherwise they are quite similar in appearance. Males may have slightly longer legs, and in the breeding season, a low fleshy keel running along the edges of the tail.
Habitat: Although found mainly in arctic and subarctic regions, this species manages to breed in small, still water bodies in a variety of habitats. These include tundra and steppe, although shady forested habitats are preferred.
Habits: In the northern part of its range activity is confined to the short summer period which lasts 3–4 months. At this time the salamanders return to small ponds or puddles for breeding. Unlike other European newts and salamanders, male Siberian Salamanders fertilise the eggs externally, after they have been released by the female. Siberian Salamanders have the remarkable ability to tolerate freezing, and adults can survive at temperatures as low as –45°C by producing 'antifreeze' chemicals in the blood and tissues. Adult salamanders leave water when breeding is finished, but stay close to the breeding pools during the winter, hibernating in thick moss or cracks and crevices in the soil.
Eggs and larvae: About 50–60 eggs are laid in two spiral-shaped sacs (about 12 cm long) which are attached to plants, stones or twigs. Unlike most other newt eggs those of the Siberian Salamander are darkly pigmented, and can survive short periods of freezing. The eggs take up to a month to complete development and from the sacs hatch larvae about 10 mm long. These have conspicuous gills and a well-developed tail fin, and grow to a size of about 45 mm before metamorphosis. The larvae appear to be able to tolerate wide fluctuations in temperature.
References: Kuzmin (1994c,d); Vorobyeva (1994).

APPENDIX

Herpetological Organisations

The organisations listed below cater for those with an interest in newts and salamanders and may be useful sources of further information. All of the organisations produce regular publications and many of them have active conservation programmes. Regular meetings and conferences provide opportunities for members to meet other herpetologists and exchange information.

The British Herpetological Society (BHS)

c/o The Zoological Society of London, Regent's Park, London NW1 4RY, UK

The largest herpetological society in Britain, the BHS has active conservation, captive breeding, education and research committees. The society produces a monthly newsletter, and the *BHS Bulletin* and the *Herpetological Journal* quarterly.

Herpetological Conservation International (HCL)

Triton House, Bramfield, Halesworth, Suffolk IP19 9AE, UK

A non-membership organisation which is primarily concerned with conservation issues in Britain and co-ordinating local herpetological groups. Publishes *Herpetofauna News* and *Herp-Line.*

Societas Europaea Herpetologica (SEH)

c/o The General Secretary, Dr M.R.K. Lambert, Natural Resources Institute, Central Avenue, Chatham Maritime, Kent ME4 4TB, UK

SEH is the main scientific herpetological organisation in Europe, and has an active conservation and mapping programme. It publishes the scientific journal *Amphibia-Reptilia*, and holds meetings every 2 years.

International Society for the Study and Conservation of Amphibians (ISSCA)

The General Secretary, ISSCA, Laboratoire des Reptiles et Amphibiens, Muséum National d'Histoire Naturelle, 25 rue Cuvier, 75005 Paris, France

This society focusses entirely on amphibians and publishes the scientific journal *Alytes.*

Declining Amphibian Population Task Force (DAPTF)

Dr John Baker, Department of Biology, The Open University, Milton Keynes MK7 6AA, UK

DAPTF is the IUCN Species Survival Commission specialist group on declining amphibian populations. It acts as a forum for research on amphibian declines, provides small project grants, and publishes the newsletter *Froglog.*

Deutsche Gesellschaft für Herpetologie und Terrarienkunde (DGHT)

DGHT-Geschäftsstelle, Postfach 14 21, D-53351 Rheinbach, Germany

A large society which caters for both amateur and professional herpetologists. Publishes the newsletter *Elaphe* and the journals *Salamandra* and *Mertensiella.*

Society for the Study of Amphibians and Reptiles (SSAR)

Department of Zoology, Ohio University, Athens, OH 45701, USA

A large society based in North America but with an international membership. Holds an annual meeting and produces a range of publications, including *Herpetological Review* and *Journal of Herpetology.*

The Herpetologist's League (HL)

Dr Rebecca Pyles, Secretary, Herpetologists's League, Dept. of Biological Sciences, Box 70726, East Tennessee State University, Johnson City, TN 37614–0726, USA

A North American-based society that publishes the journal *Herpetologica* which has an international coverage.

American Society of Ichthyologists and Herpetologists (ASIH)

ASIH Business Office, Texas Natural History Collection, R4000, University of Texas, Austin, Texas 78705–1100, USA.

A large society catering for those interested in reptiles, amphibians and fishes. Members' articles are published in the scientific journal *Copeia.*

References

Accordi, F., Massarek, A. & Nobili, G. (1990) Ecological responses in a population of smooth newts (*Triturus vulgaris meridionalis*) in an unpredictable environment. *Herpetol. J.* **1**: 509–513.

Ahlberg, P.E. & Milner, A.R. (1994) The origin and early diversification of tetrapods. *Nature* **368**: 507–514.

Alcher, M. (1980) Contribution a l'étude de développement de 'Urodèle *Euproctus platycephalus* (Gravenhorst, 1829). *Vie Milieu* **30**: 157–164.

Alcher, M. (1981a) Sur l'existence de soins parentaux chez *Euproctus montanus* (Urodela, Salamandridae). *Amphibia-Reptilia* **2**: 189–194.

Alcher, M. (1981b) Reproduction en élevage de *Euproctus platycephalus* (Urodela, Salamandridae). *Amphibia-Reptilia* **2**: 97–105.

Alcher, M. (1988) Durée du développment larvaire de l'Urodèle *Euproctus montanus* (Amphibia, Salamandridae) dans deux localitiés corses d'altitudes différentes. *Alytes* 7: 63–74.

Alcobendas, M., Dopazo, H. & Alberch, P. (1994) Genetic structure and differentiation in *Salamandra salamandra* populations from the Northern Iberian peninsula. *Mertensiella* **4**: 7–23.

Aljancic, M., Bulog, B., Kranjc, A., Josipovic, D., Sket, B. & Skoberne, P. (1993) *Proteus. The Mysterious Ruler of Karst Darkness*, Ljublijana, Slovenia: Vitrum.

Andreone, F. (1985) Observations on the defensive behaviour of the crested newt (*Triturus cristatus carnifex*). *Br. Herpetol. Soc. Bull.* **13**: 37–38.

Andreone, F. (1992) Observations on the territorial and reproductive behaviour of *Salamandra lanzai* and considerations about its protection (Amphibia: Salamandridae). *Br. Herpetol. Soc. Bull.* **39**: 31–33.

Andreone, F. & Dore, B. (1992) Adaptation of the reproductive cycle in *Triturus alpestris apuanus* to an unpredictable habitat. *Amphibia-Reptilia* **13**: 251–261.

Andreone, F. & Giacoma, C. (1989) Breeding dynamics of *Triturus carnifex* at a pond in northwestern Italy (Amphibia, Urodela, Salamandridae). *Hol. Ecol.* **12**: 219–223.

Arano, B. & Arntzen, J.W. (1987) Genetic differentiation in the alpine newt, *Triturus alpestris*. In van Gelder, J.J., Strijbosch, H. & Bergers, P.J.M. (eds), *Proceedings 4th O.G.M. of Societas Europaea Herpetologica*, pp. 21–24, Nijmegen: Faculty of Sciences.

Arnold, E.N. & Burton, J.A. (1978) *A Field Guide to the Reptiles and Amphibians of Britain and Europe*, London: Collins.

Arnold, S.J. (1977) The evolution of courtship behavior in New World salamanders with some notes on Old World salamandrids. In Taylor, D.H. & Guttman, S.I. (eds), *The Reproductive Biology of Amphibians*, pp. 141–183, New York: Plenum.

Arnold, S.J. (1987) The comparative ethology of courtship in salamandrid salamanders, 1. *Salamandra* and *Chioglossa*. *Ethology* 74: 133–145.

Arntzen, J.W. (1981) Ecological observations on *Chioglossa lusitanica* (Caudata, Salamandridae). *Amphibia-Reptilia* **3/4**: 187–203.

Arntzen, J.W. & de Wijer, P. (1989) On the distribution of the Palaearctic newts (genus *Triturus*) including the description of a five species pond in western France. *Br. Herpetol. Soc. Bull.* **30**: 6–11.

Arntzen, J.W. & Hedlund, L. (1990) Fecundity of the newts *Triturus cristatus, T. marmoratus* and their natural hybrids in relation to species coexistence. *Hol. Ecol.* **13**: 325–332.

Arntzen, J.W. & Sparreboom, M. (1989) A phylogeny for the Old World newts, genus *Triturus:* biochemical and behavioural data. *J. Zool.* **219**: 645–664.

Arntzen, J.W. & Teunis, S.F.M. (1993) A six year study on the population dynamics of the crested newt (*Triturus cristatus*) following the colonization of a newly created pond. *Herpetol. J.* **3**: 99–110.

Arntzen, J.W. & Wallis, G.P. (1994) The 'WOLTERSTORFF Index' and its value to the taxonomy of the crested newt superspecies. *Abh. Ber. Nat.* **17**: 57–66.

Aronsson, S. & Stenson, J.A.E. (1995) Newt–fish interactions in a small forest lake. *Amphibia-Reptilia* **16**: 177–184.

Baker, J.M.R. (1990a) Body size and spermatophore production in the smooth newt (*Triturus vulgaris*). *Amphibia-Reptilia* **11**: 173–184.

Baker, J.M.R. (1990b) Growth of larval and juvenile newts. PhD thesis, Milton Keynes: The Open University.

Baker, J.M.R. (1992) Egg production in the smooth newt (*Triturus vulgaris*). *Herpetol. J.* **2**: 90–93.

Baker, J.M.R. (1994) Survival of some captive-reared crested newts on release into the wild. *Br. Herpetol. Soc. Bull.* **49**: 11–13.

Ballasina, D. (1984) *Amphibians of Europe. A Colour Field Guide*, Newton Abbot: David & Charles.

Banks, B. & Laverick, G.L. (1986) Garden ponds as amphibian breeding sites in a conurbation in the north east of England (Sunderland, Tyne and Wear). *Herpetol. J.* **1**: 44–50.

Baran, I. & Ücüncü, S. (1994) The state of *Mertensiella luschani* in Turkey. *Mertensiella* **4**: 33–40.

Baylis, H.A. (1939) Delayed reproduction in the spotted salamander. *Proc. Zool. Soc. London* **109**: 243–246.

Beattie, R.C. (1985) The date of spawning in populations of the Common frog (*Rana temporaria*) from different altitudes in northern England. *J. Zool.* (A) **205**: 137–154.

Beebee, T.J.C. (1995) Amphibian breeding and climate. *Nature* **374**: 219–220.

Bell, G. (1977) The life of the smooth newt (*Triturus vulgaris*) after metamorphosis. *Ecol. Monogr.* **47**: 279–299.

Bell, G. (1979) Populations of crested newts, *Triturus cristatus*, in Oxfordshire, England. *Copeia* 1979, 350–353.

Bell, G. & Lawton, J.H. (1975) The ecology of the eggs and larvae of the smooth newt (*Triturus vulgaris* (Linn.)). *J. Anim. Ecol.* **44**: 393–424.

Benson, P. (1979) Observations of gills on the adult smooth newt. *Herptile* **4**: 33–34.

Bentley, P.J. (1966) The physiology of the urinary bladder of Amphibia. *Biol. Rev.* **41**: 275–316.

Blab, J. & Blab, L. (1981) Quantitative Analysen zur Phänologie, Erfaßbarkeit und Populationsdynamik von Molchbeständen des Kottenforstes bei Bonn (Amphibia: Caudata: Salamandridae). *Salamandra* **17**: 147–172.

Bolkay, S. J. (1928) Die Schädel der Salamandrinen, mit besonderer Rücksicht auf ihre systematische Bedeutung. *Z. Anat. Entwicklungsgesch.* **86**: 259–319.

Boutilier, R.G., Stiffler, D.F. & Toews, D.P. (1992) Exchange of respiratory gases, ions, and water in amphibious and aquatic amphibians. In Feder, M.E. & Burggren, W.W. (eds), *Environmental Physiology of the Amphibians*, pp. 81–124, Chicago: University of Chicago Press.

Brattstrom, B.H. (1979) Amphibian temperature regulation studies in the field and laboratory. *Am. Zool.* **19**: 345–356.

Breuil, M. & Guillaume, C-P. (1985) Etude de quelques populations de Tritons alpestres néoténiques (*Triturus alpestris*) (Amphibia Caudata Salamandridae) du sud de la Yougoslavie. *Bull. Soc. Zool. Fr.* **109**: 377–389.

Briegleb, W. (1962) Zur Biologie und Ökologie des Grottenolms (*Proteus anguinus* Laur., 1768). *Z. Morphol. Ökol. Tiere* **51**: 271–334.
Brodie, E.D. Jr. (1983) Antipredator adaptations of salamanders: evolution and convergence among terrestrial species. In Margaris, N.S., Arianoutsou-Faraggitaki, M. & Reitner, R.J. (eds), *Plant, Animal, and Microbial Adaptations to Terrestrial Environment*, pp. 109–133, New York: Plenum.
Brodie, E.D. Jr. & Smatresk, N.J. (1990) The antipredator arsenal of fire salamanders: spraying of secretions from highly pressurized dorsal skin glands. *Herpetologica* **46**: 1–7.
Bucci-Innocenti, S., Ragghianti, M. & Mancino, G. (1983) Investigation of karyology and hybrids in *Triturus boscai* and *T. vittatus*, with a reinterpretation of the species groups within *Triturus* (Caudata: Salamandridae). *Copeia* 1983, 662–672.
Busack, S.D. (1976) A review of the biology of the gold-striped salamander, *Chioglossa lusitanica* (Amphibia: Salamandridae). *Biol. Conserv.* **10**: 309–319.
Caccone, A., Milinkovitch, M.C., Sbordoni, V. & Powell, J.R. (1994) Molecular biogeography: using the Corsica–Sardinia microplate disjunction to calibrate mitochondrial rDNA evolutionary rates in mountain newts (*Euproctus*). *J. Evol. Biol.* 7: 227–245.
Clergue-Gazeau, M. (1965) Étude comparative de l'euprocte des lacs et de l'euprocte cavernicole. *Ann. Speleol.* **20**: 301–316.
Cogalniceanu, D. (1994) The relative importance of vision and olfaction in mate recognition in male newts (genus *Triturus*). *Herpetologica* **50**: 344–349.
Cooke, A.S. & Frazer, J.F.D. (1976) Characteristics of newt breeding sites. *J. Zool.* **178**: 223–236.
Corbett, K. (1989) *Conservation of European Reptiles and Amphibians*, London: Christopher Helm.
Crews, D. (1987) Diversity and evolution of behavioral controlling mechanisms. In Crews, D. (ed.), *Psychobiology of Reproductive Behavior. An Evolutionary Perspective*, pp. 88–119, New Jersey: Prentice-Hall.
De Fonseca, P.H. & Joqué, R. (1982) The palmate newt *Triturus helveticus* (Raz.) in Flanders (Belgium). Distribution and habitat preferences. *Biol. Conserv.* **23**: 297–307.
Degani, G. (1982) Temperature tolerance in three populations of salamanders, *Salamandra salamandra* L. *Br. J. Herpetol.* **6**: 186–187.
Degani, G. (1984) Temperature selection in *Salamandra salamandra* (L.) larvae and juveniles from different habitats. *Biol. Behav.* **9**: 175–183.
Degani, G. & Mendelssohn, H. (1982) Seasonal activity of *Salamandra salamandra* (L.) (Amphibia: Urodela: Salamandridae) in the headwaters of the Jordan River. *Isr. J. Zool.* **31**: 77–85.
Degani, G. & Mendelssohn, H. (1983) The habitats, distribution and life history of *Triturus vittatus vittatus* (Jenyns) in the Mount Meron area (Upper Galilee, Israel). *Br. J. Herpetol.* **6**: 317–319.
Degani G. & Warburg, M.R. (1980) The response to substrate moisture of juvenile and adult *Salamandra salamandra* (L.) (Amphibia; Urodela). *Biol. Behav.* **5**: 281–290.
Denton, J. (1990) Defensive reflexes in newts of the genus *Triturus*. *Br. Herpetol. Soc. Bull.* **32**: 30.
Diaz-Paniagua, C. (1989) Oviposition behavior of *Triturus marmoratus pygmaeus*. *J. Herpetol.* **23**: 159–163.
Diaz-Paniagua, C. (1992) Variability in timing of larval season in an amphibian community in SW Spain. *Ecography* **15**: 267–272.
Dolmen, D. (1980) Distribution and habitat of the smooth newt, *Triturus vulgaris* (L.), and the warty newt, *T. cristatus* (Laurenti), in Norway. In Coborn, J. (ed.), *Proceedings European Herpetological Symposium*, pp. 127–139, Oxford: Cotswold Wildlife Park.
Dolmen, D. (1983) Diel rhythms and microhabitat preference of the newts *Triturus vulgaris* and *T. cristatus* at the northern border of their distribution. *J. Herpetol.* **17**: 23–31.
Dolmen, D. (1987) Hazards to Norwegian amphibians. In van Gelder, J.J., Strijbosch, H. & Bergers, P.J.M. (eds), *Proceedings 4th O.G.M. of Societas Europaea Herpetologica*, pp. 119–122, Nijmegen: Faculty of Sciences.

Dolmen, D. & Koksvik, J.I. (1983) Food and feeding habits of *Triturus vulgaris* (L.) and *T. cristatus* (Laurenti) (Amphibia) in two bog tarns in central Norway. *Amphibia-Reptilia* **4**: 17–24.

Dorazi, R., Chesnel, A. & Dournon, C. (1995) Opposite sex determination of gonads in two *Pleurodeles* species may be due to a temperature-dependent inactivation of sex chromosomes. *J. Hered.* **86**: 28–31.

Duellman, W.E. & Trueb, L. (1986) *Biology of Amphibians*, New York: McGraw-Hill.

Durand, J. & Bouillon, M. (1964) Observations sur le développement du protée, *Proteus anguinus* Laurenti. *C. R. Acad. Sci. Paris* **259**: 4801–4804.

Dzukic, G., Kalezic, M.L., Trvtkovic, N. & Djorovic, A. (1990) An overview of the occurrence of paedomorphosis in Yugoslav newt (*Triturus*, Salamandridae) populations. *Br. Herpetol. Soc. Bull.* **34**: 16–22.

Epperlein, H.H. & Junginger, M. (1982) The normal development of the newt, *Triturus alpestris* (Daudin). *Amphibia-Reptilia* **2**: 295–308.

Faria, M.M. (1993) Sexual behaviour of Bosca's newt, *Triturus boscai*. *Amphibia-Reptilia* **14**: 169–185.

Faria, M.M. (1995) A field study of reproductive interactions in Bosca's newt. *Triturus boscai*. *Amphibia-Reptilia* **16**: 357–374.

Fog, K. (1995). Amphibian conservation in Denmark. *Froglog* **13**: 1–2.

Francillon-Vieillot, H., Arntzen, J.W. & Géraudie, J. (1990) Age, growth and longevity of sympatric *Triturus cristatus*, *T. marmoratus* and their hybrids (Amphibia, Urodela): a skeletochronological comparison. *J. Herpetol.* **24**: 13–22.

Gallien, L. & Bidaud, O. (1959) Table chronologique de développement chez *Triturus helveticus* Razoumowsky. *Bull. Soc. Zool. Fr.* **84**: 22–32.

Garcia-Paris, M. (1985). *Los Anfibios de España*, Madrid: Ministerio de Agricultura, Pesca y Alimentacion.

Garcia-Paris, M. & Deban, S. M. (1995) A novel antipredator mechanism in salamanders: rolling escape in *Hydromantes platycephalus*. *J. Herpetol.* **29**: 149–151.

Garcia-Paris, M., Herrero, P., Martin, C., Dorda, J., Esteban M. & Arano, B. (1993) Morphological characterization, cytogenetic analysis, and geographical distribution of the Pygmy Marbled Newt, *Triturus marmoratus pygmaeus* (Wolterstorff, 1905) (Caudata: Salamandridae). *Bijdr. Dierkd.* **63**: 3–14.

Geffen, E., Gafny, S. & Gasith, A. (1987) Contribution to the knowledge of the biology of the banded newt (*Triturus vittatus vittatus*) in rainpools in Israel. *Isr. J. Zool.* **34**: 213–223.

Giacoma, C. & Balletto, E. (1988) Phylogeny of the salamandrid genus *Triturus*. *Boll. Zool.* **55**: 337–360.

Giacoma, C. & Sparreboom, M. (1987) On the sexual behaviour of the Italian newt, *Triturus italicus* (Caudata: Salamandridae). *Bijdr. Dierkd.* **57**: 19–30.

Giesenberg, A. (1991) Untersuchungen zur Entwicklung der Larven von *Salamandra salamandra terrestris* Lacépède, 1788 aus dem Landkreis Harburg (Niedersachsen). *Salamandra* **27**: 97–107.

Gislen, T. & Kauri, H. (1959) Zoogeography of the Swedish amphibians and reptiles. *Acta Vert.* **1**: 193–397.

Glandt, D. (1980) Die quantitative Vertikalverbreitung der Molch-Arten, Gattung *Triturus* (Amphibia, Urodela), in der Bundesrepublik Deutschland. *Bonn. Zool. Beitr.* **31**: 97–110.

Glandt, D. (1982) Abundanzmessungen an mitteleuropäischen *Triturus*-Populationen (Amphibia, Salamandridae). *Amphibia-Reptilia* **3**: 317–326.

Green, A.J. (1989) The sexual behaviour of the great crested newt, *Triturus cristatus* (Amphibia: Salamandridae). *Ethology* **83**: 129–153.

Griffiths, R.A. (1981) Physical abnormalities and accessory limb growth in the smooth newt, *Triturus vulgaris*. *Br. J. Herpetol.* **6**: 180–182.

Griffiths, R.A. (1984) Seasonal behaviour and intrahabitat movements in an urban population of Smooth newts, *Triturus vulgaris* (Amphibia: Salamandridae). *J. Zool.* **203**: 241–251.

Griffiths, R.A. (1985) Diel profile of behaviour in the smooth newt, *Triturus vulgaris* (L.): an analysis of environmental cues and endogenous timing. *Anim. Behav.* **33**: 573–582.

Griffiths, R.A. (1986) Feeding niche overlap and food selection in smooth and palmate newts, *Triturus vulgaris* and *T. helveticus*, at a pond in mid-Wales. *J. Anim. Ecol.* **55**: 201–214.

Griffiths, R.A. (1987) Microhabitat and seasonal niche dynamics of smooth and palmate newts, *Triturus vulgaris* and *T. helveticus*, at a pond in mid-Wales. *J. Anim. Ecol.* **56**: 441–451.

Griffiths, R.A. & Mylotte, V.J. (1987) Microhabitat selection and feeding relations of smooth and warty newts, *Triturus vulgaris* and *T. cristatus*, at an upland pond in mid-Wales. *Hol. Ecol.* **10**: 1–7.

Griffiths, R.A. & de Wijer, P. (1994) Differential effects of pH and temperature on embryonic development in the British newts (*Triturus*). *J. Zool.* **234**: 613–622.

Griffiths, R.A., Harrison, J.D. & Gittins, S.P. (1986) The breeding migrations of amphibians at Llysdinam pond, Wales: 1981–1985. In Rocek, Z. (ed.), *Studies in Herpetology*, pp. 543–546, Prague: Charles University.

Griffiths, R.A., Roberts, J.M. & Sims, S. (1987) A natural hybrid newt, *Triturus helveticus* × *T. vulgaris*, from a pond in mid-Wales. *J. Zool.* **213**: 133–140.

Griffiths, R.A., de Wijer, P. & Brady, L. (1993) The effect of pH on the development of eggs and larvae of Smooth and Palmate newts, *Triturus vulgaris* and *T. helveticus*. *J. Zool.* **230**: 401–409.

Griffiths, R.A., de Wijer, P. & May, R.T. (1994) Predation and competition within an assemblage of larval newts (*Triturus*). *Ecography* **17**: 176–181.

Grossenbacher, K. (1994) Zur Systematik und Verbreitung der Alpensalamander (*Salamandra atra atra, Salamandra atra aurorae, Salamandra lanzai*). *Abh. Ber. Naturkunde* **17**: 75–81.

Grüsser-Cornehls, U. & Himstedt, W. (1976) The urodele visual system. In Fite, K.V. (ed.), *The Amphibian Visual System – A Multidisciplinary Approach*, pp. 203–266, New York: Academic Press.

Guerrero, F., Pérez-Mellado, V., Gil, M.J. & Lizana, M. (1990) Food habits and trophic availability in the high mountain population of the spotted salamander from Spain (*Salamandra salamandra almanzoris*) (Caudata: Salamandridae). *Folia Zool.* **39**: 341–353.

Habermehl, G. (1994) The biological relevance of *Salamandra* venom. *Mertensiella* **4**: 209–214.

Häfeli, H. (1971) Zur Fortpflanzungsbiologie des Alpensalamanders (*Salamandra atra* Laur.). *Rev. Suisse Zool.* **78**: 235–293.

Hagström, T. (1979). Population ecology of *Triturus cristatus* and *T. vulgaris* (Urodela) in SW Sweden. *Hol. Ecol.* **2**: 108–114.

Hairston, N.G. & Wiley, R.H. (1993) No decline in salamander (Amphibia: Caudata) populations: a twenty-year study in the southern Appalachians. *Brimleyana* **18**: 59–64.

Halliday, T.R. (1974) Sexual behavior of the smooth newt, *Triturus vulgaris* (Urodela: Salamandridae). *J. Herpetol.* **8**: 277–292.

Halliday, T.R. (1977) The courtship of European newts. An evolutionary perspective. In Taylor, D.H. & Guttman, S.I. (eds), *The Reproductive Biology of Amphibians*, pp. 185–232, New York: Plenum.

Halliday, T.R. (1990) The evolution of courtship behavior in newts and salamanders. *Adv. Study Behav.* **19**: 137–169.

Halliday, T.R. & Sweatman, H.P.A. (1976) To breathe or not to breathe; the newt's problem. *Anim. Behav.* **24**: 551–561.

Harrison, J.D., Gittins, S.P. & Slater, F.M. (1983) The breeding migrations of Smooth and Palmate newts (*Triturus vulgaris* and *T. helveticus*) at a pond in mid-Wales. *J. Zool.* **199**: 249–258.

Hedlund, L. & Robertson, J.G.M. (1989) Lekking behaviour in crested newts, *Triturus cristatus*. *Ethology* **80**: 111–119.

Heyer, W.R., Donnelly, M.A., McDiarmid, R.W., Hayek. L.-A. C. & Foster, M.S. (1994) *Measuring and Monitoring Biological Diversity. Standard Methods for Amphibians*, Washington: Smithsonian Institution Press.

Hilton-Brown, D. & Oldham, R.S. (1991) The status of the widespread amphibians and reptiles in Britain, 1990, and changes during the 1980's. Nature Conservancy Council Contract Survey No. 131, Peterborough: NCC.

Himstedt, W. (1965) Beobachtungen zum Paarungsverhalten des Feursalamanders (*Salamandra salamandra* L.). *Zool. Anz.* **175**: 295–300.

Himstedt, W. (1967) Experimentelle Analyse der optischen Sinnesleistungen im Beutefangverhalten der einheimischen Urodelen. *Zool. Jahrb. Abt. Allg. Zool. Physiol. Tiere* **73**: 281–320.

Himstedt, W. (1971) Die Tagesperiodik von Salamandriden. *Oecologia* **8**: 194–208.

Himstedt, W. (1973) Die spektrale Empfindlichkeit von Urodelen in Abhängigkeit von Metamorphose, Jahreszeit und Lebensraum. *Zool. Jahrb. Abt. Allg. Zool. Physiol. Tiere* **77**: 246–274.

Himstedt, W. (1979) The significance of color signals in partner recognition of the newt *Triturus alpestris. Copeia* **1979**: 40–43.

Himstedt, W. (1994) Sensory systems and orientation in *Salamandra salamandra. Mertensiella* **4**: 225–239.

Himstedt, W., Kopp, J. & Schmidt, W. (1982) Electroreception guides feeding behavior in amphibians. *Naturwissenschaften* **69**: 552.

Horner, H.A. & Macgregor, H.C. (1985) Normal development in newts (*Triturus*) and its arrest as a consequence of an unusual chromosome situation. *J. Herpetol.* **19**: 261–270.

Hutchison, V.H. & Dupré, R.K. (1992) Thermoregulation. In Feder, M.E. & Burggren, W.W. (eds), *Environmental Physiology of the Amphibians*, pp. 206–249. Chicago: University of Chicago Press.

Joly, J. (1960) La conservation des spermatozoides et les particularités histophysiologiques du réceptacle séminale chez la salamandre *Salamandra salamandra taeniata. C. R. Acad. Sci. Paris* **250**: 2269–2271.

Joly, J. (1961) Le cycle sexuel biennial chez la femelle salamandre tachetée *Salamandra salamandra quadri-virgata* dans les hautes-Pyrénées. *C. R. Acad. Sci. Paris* **252**: 3145–3147.

Joly, J. (1966) Sur l'éthologie sexuelle de *Salamandra salamandra. Z. Tierpsychol.* **23**: 8–27.

Joly, J. & Lardry, M. (1964) Électricité atmosphérique et activité dans la nature d'un Amphibien Urodèle: *Salamandra salamandra. C. R. Acad. Sci., Paris* **259**: 2920–2922.

Joly, J., Chesnel, F. & Boujard, D. (1994) Biological adaptations and reproductive strategies in the genus *Salamandra. Mertensiella* **4**: 255–269.

Joly, P. & Giacoma, C. (1992) Limitation of similarity and feeding habits in three syntopic species of newts (*Triturus*, Amphibia). *Ecography* **15**: 401–411.

Joly, P. & Miaud, C. (1993) How does a newt find its pond? The role of chemical cues in migrating newts (*Triturus alpestris*). *Ethol. Ecol. Evol.* **5**: 447–455.

Kalezic, M.L., Cvetkovic, D., Djorovic, A. & Dzukic, G. (1994) Paedomorphosis and differences in life-history traits of two neighbouring crested newt (*Triturus carnifex*) populations. *Herpetol. J.* **4**: 151–158.

Kastle, W. (1986) Rival combats in *Salamandra salamandra.* In: Rocek, Z. (ed.), *Studies in Herpetology*, pp. 525–528, Prague: Charles University.

Klewen, R. (1986) Population ecology of *Salamandra salamandra terrestris* in an isolated habitat. In Rocek, Z. (ed.), *Studies in Herpetology*, pp. 395–398, Prague: Charles University.

Klewen, R. (1991) *Die Landsalamander Europas I*, Wittenberg Lutherstadt: Ziemsen.

Kuzmin, S.L. (1994a) The problem of declining amphibian populations in the Commonwealth of Independent States and adjacent territories. *Alytes* **12**: 123–134.

Kuzmin, S.L. (1994b) Commercial collecting as a threat for amphibian and reptile species of the former Soviet Union. *Species* **23**: 47–48.

Kuzmin, S.L. (1994c) Bibliography of the Siberian newt (*Salamandrella keyserlingii* Dybowski, 1870). Smithsonian Herpetological Information Service No. 102.

Kuzmin, S.L. (1994d) The geographical range of *Salamandrella keyserlingii* Dybowski, 1870: ecological and historical implications. *Abh. Ber. Naturkd.* **17**: 177–183.

Kuzmin, S.L. & Meschersky, I.G. (1987) Feeding dynamics of the smooth newt (*Triturus vulgaris*) during the ontogenesis. *Zool. Zh.* **66**: 75–84 (in Russian).

Laan, R. & Verboom, B. (1990) Effects of pool size and isolation on amphibian communities. *Biol. Conserv.* **54**: 251–262.

Lanza, B. & Corti, C. (1993) Erpetofauna Italiana: 'Aquisizioni' ed estinzioni nel corso del Novecento. *Suppl. Ric. Biol. Selv.* **21**: 5–49.

Lanza, B., Nascetti, G. & Bullini, L. (1986) A new species of *Hydromantes* from eastern Sardinia and its genetic relationships with the other Sardinian plethodontids (Amphibia: Urodela). *Boll. Mus. Reg. Sci. Nat. Torino* **4**: 261–269.

Lengvenus, W. & Parzefall, J. (1992) The role of the visual reaction in the behaviour of an epigean and a cave living population of *Euproctus asper* Duges (Salamandridae, Urodela). *Mem. Biospéol.* **19**: 111–118.

Littleford, R.A., Keller, W.F. & Phillips, N.E. (1947) Studies on the vital limits of water loss in the plethodontid salamanders. *Ecology* **28**: 440–447.

Luthardt, G. & Roth, G. (1983) The interaction of the visual and the olfactory systems in guiding prey catching behaviour in *Salamandra salamandra* (L.). *Behaviour* **83**: 69–79.

Luthardt-Laimer, G. (1983) Ontogeny of preferences to visual prey stimulus parameters in salamanders. *J. Herpetol.* **17**: 221–227.

Luthardt-Laimer, G. & Roth, G. (1983) Reduction of visual inhibition to stationary prey by early experience in *Salamandra salamandra* (L.). *Z. Tierpsychol.* **63**: 294–302.

Macgregor, H.C. & Horner, H.A. (1980) Heteromorphism for chromosome 1, a requirement for normal development in crested newts. *Chromosoma* **76**: 111–122.

Macgregor, H.C., Sessions, S.K. & Arntzen, J.W. (1990) An integrative analysis of phylogenetic relationships among newts of the genus *Triturus* (family Salamandridae), using comparative biochemistry, cytogenetics and reproductive interactions. *J. Evol. Biol.* **3**: 329–373.

Mann, W., Dorn, P. & Brandl, R. (1991) Local distribution of amphibians: the importance of habitat fragmentation. *Global Ecol. Biogeogr. Lett.* **1**: 36–41.

Martens, H. & Jelden, D. (1992) Pet trade in reptiles and amphibians – perspectives of their conservation status within the European Community (EC). In Korsós, Z. & Kiss, I. (eds). *Proceedings 6th O.G.M. of Societas Europaea Herpetologica*, pp. 309–313, Budapest: Hungarian Natural History Museum.

May, R. (1993) Factors affecting the growth and survival of the smooth and palmate newt (*Triturus vulgaris* and *T. helveticus*) with particular attention to acidification and temperature. BSc (Tech.) Biology Professional Training Report. DICE, University of Kent.

McLee, A.G. & Scaife, R.W. (1993) The colonisation by great crested newts (*Triturus cristatus*) of a water body following treatment with a piscicide to remove a large population of sticklebacks (*Gasterosteus aculeatus*). *Br. Herpetol. Soc. Bull.* **42**: 6–9.

Miaud, C. (1993) Predation on newt eggs (*Triturus alpestris* and *T. helveticus*): identification of predators and protective role of oviposition behaviour. *J. Zool.* **231**: 575–582.

Miaud, C. (1994) Role of wrapping behavior on egg survival in three species of *Triturus* (Amphibia: Urodela). *Copeia* **1994**: 535–537.

Miller, M.R. & Robbins, M.E. (1954) The reproductive cycle in *Taricha torosa*. *J. Exp. Zool.* **125**: 415–443.

Milner, A.C., Milner, A.R. & Estes, R. (1982) Amphibians and squamates from the Upper Eocene of Hordle Cliff, Hampshire – a preliminary report. *Tert. Res.* **4**: 149–154.

Milner, A.R. (1988) The relationships and origins of living amphibians. In Benton, M.J. (ed.), *The Phylogeny and Classification of the Tetrapods, Vol. 1: Amphibians, Reptiles, Birds*, Systematics Association Special Volume no. 35A, pp. 59–102, Oxford: Clarendon Press.

Milner, A.R. (1993) The Paleozoic relatives of lissamphibians. *Herpetol. Monogr.* 7: 8–27.

Möller, A. (1951) Die Struktur des Auges bei Urodelen verschiedener Körpergröße. *Zool. Jahrb. Abt. Allg. Zool. Physiol. Tiere* **62**: 138–182.

Nardi, I. (1991) Cytogenetics of the European plethodontid salamanders, genus *Hydromantes.* In Green, D.M. & Sessions, S.K. (eds), *Amphibian Cytogenetics and Evolution,* pp. 131–156, San Diego: Academic Press.

Nascetti, G., Andreone, F., Capula, M. & Bullini, L. (1988) A new *Salamandra* species from south-western Alps. *Boll. Mus. Reg. Sci. Nat. Torino* **6**: 617–638.

Nöllert, A. & Nöllert, C. (1992) *Die Amphibien Europas,* Stuttgart: Franckh-Kosmos.

Nuutinen, V. & Ranta, E. (1986) Size-selective predation on zooplankton by the smooth newt, *Triturus vulgaris. Oikos* **47**: 83–91.

Oldham, R.S., Musson, S. & Humphries, R.N. (1991) Translocation of crested newt populations in the U.K. *Herpetofauna News* **2**(5): 3–5.

Oosterbroek, P. & Arntzen, J. W. (1992) Area-cladograms of circum-Mediterranean taxa in relation to Mediterranean palaeogeography. *J. Biogeogr.* **19**: 3–20.

Ormerod, E.A. (1872) Observations on the cutaneous exudation of the *Triton cristatus* or great Water Newt. *J. Linn. Soc.* **11**: 493–496.

Özeti, N. (1979) Reproductive biology of the salamander *Mertensiella luschani antalyana. Herpetologica* **35**: 193–197.

Pechmann, J.H.K., Scott, D.E., Gibbons, J.W., Semlitsch, R.D. & Caldwell, J.P. (1991) Declining amphibian populations: the problem of separating human impacts from natural fluctuations. *Science* **253**: 892–895.

Pecio, A. (1992) Insemination and egg laying dynamics in the smooth newt, *Triturus vulgaris,* in the laboratory. *Herpetol. J.* **2**: 5–7.

Pecio, A. & Rafinski, J. (1985) Sexual behaviour of the Montandon's newt, *Triturus montandoni* (Boulenger) (Caudata: Salamandridae). *Amphibia-Reptilia* **6**: 11–22.

Plasa, L. (1979) Heimfindeverhalten bei *Salamandra salamandra* (L.). *Z. Tierpsychol.* **51**: 113–125.

Pocrnjic, S. (1979) Thyroid activity and its significance for neotenia and thermal adaptation of oxygen consumption in the alpine newt. *Arh. Biol. Nauka, Beograd* **20**: 105–117.

Polymeni, R. (1994) On the biology of *Mertensiella luschani* (Steindachner, 1891): a review. *Mertensiella* **4**: 301–314.

Radovanovic, M. (1951) A new race of the Alpine Newt from Yugoslavia. *Br. J. Herpetol.* **1**: 93–97.

Rafinski, J. & Pecio, A. (1992) The courtship behaviour of the Bosca's newt, *Triturus boscai* (Amphibia: Salamandridae). *Folia Biol. (Kraków)* **40**: 155–165.

Ranta, E. & Nuutinen, V. (1985) Foraging by the smooth newt (*Triturus vulgaris*) on zooplankton: functional response and diet choice. *J. Anim. Ecol.* **54**: 275–293.

Ranta, E. McManus, J. & Leikola, N. (1990) Non-visual detection of prey patches by the smooth newt (*Triturus vulgaris*). *J. Herpetol.* **24**: 202–204.

Raxworthy, C.J. (1989) Courtship, fighting and sexual dimorphism of the banded newt *Triturus vittatus ophryticus. Ethology* **81**: 148–170.

Raxworthy, C.J. (1990) A review of the smooth newt (*Triturus vulgaris*) subspecies, including an identification key. *Herpetol. J.* **1**: 481–492.

Roth, G. (1987) *Visual Behaviour in Salamanders,* Berlin: Springer-Verlag.

Salthe, S.N. (1963) The egg capsules in the Amphibia. *J. Morphol.* **13**: 161–171.

Salthe, S.N. (1969) Reproductive modes and the numbers and sizes of ova in the urodeles. *Am. Midl. Nat.* **81**: 467–490.

Salvidio, S. (1993) Life history of the European plethodontid salamander *Speleomantes ambrosii* (Amphibia, Caudata). *Herpetol. J.* **3**: 55–59.

Salvidio, S., Lattes, A., Tavano, M., Melodia, F. & Pastorino, M. V. (1994) Ecology of a *Speleomantes ambrosii* population inhabiting an artificial tunnel. *Amphibia-Reptilia* **15**: 35–45.

Schabetsberger, R. (1993) Der Bergmolch (Triturus alpestris, LAURENTI) als Endkonsument in einem alpinen Kartsee (Dreibrüdersee, 1643 m, Totes Gebirge). Doctoral Dissertation, University of Salzburg, Austria.

Schabetsberger, R. & Jersabek, C.D. (1995) Alpine newts (*Triturus alpestris*) as top predators in a high-altitude karst lake: daily food consumption and impact on the copepod *Arctodiaptomus alpinus. Freshwater Biol.* **33**: 47–61.

Schaeffer, B. (1941) The morphological and functional evolution of the tarsus in amphibians and reptiles. *Bull. Am. Mus. Nat. Hist.* **78**: 395–472.

Schmidtler, J.J. & Schmidtler, J.F. (1983) Verbreitung, Ökologie und innerartliche Gliederung von *Triturus vulgaris* in den adriatischen Küstengebieten. *Spixiana* **6**: 229–249.

Scholz, K.P. (1993) Untersuchungen zur Phylogenie der Familie Salamandridae. Diplomarbeit am Fachbereich Biologie der Phillipps-Universität Marburg.

Schoorl, J. & Zuiderwijk, A. (1981) Ecological isolation in *Triturus cristatus* and *Triturus marmoratus* (Amphibia: Salamandridae). *Amphibia-Reptilia* **1**: 235–252.

Schultschik, G. (1994) Zur Fortpflanzungsbiologie von *Mertensiella caucasica* (Waga, 1876) (Urodela: Salamandridae). *Abh. Ber. Naturkd.* **17**: 163–175.

Shoemaker, V.H., Hillman, S.S., Hillyard, S.D., Jackson, D.C., McClanahan, L.L., Withers, P.C. & Wygoda, M.L. (1992) Exchange of water, ions, and respiratory gases in terrestrial amphibians. In Feder, M.E. & Burggren, W.W. (eds), *Environmental Physiology of the Amphibians*, pp. 125–150. Chicago: University of Chicago Press.

Sket, B. & Arntzen, J.W. (1994) A black, non-troglomorphic amphibian from the karst of Slovenia: *Proteus anguinus parkelj* n. ssp. (Urodela: Proteidae). *Bijdr. Dierkd.* **64**: 33–53.

Smirina, E. & Rocek, Z. (1976) On the possibility of using annual bone layers of alpine newts, *Triturus alpestris* (Amphibia: Urodela), for their age determination. *Vestn. Cesk. Spol. Zool.* **40**: 232–237.

Sparreboom, M. & Arntzen, J.W. (1987) A survey of behaviour in the Old World newts (genus *Triturus*). In van Gelder, J.J., Strijbosch, H. & Bergers, P.J.M. (eds), *Proceedings 4th O.G.M. of Societas Europaea Herpetologica*, pp. 369–372, Nijmegen: Faculty of Sciences.

Spight, T.M. (1968) The water economy of salamanders: evaporative water loss. *Physiol. Zool.* **41**: 195–203.

Stenhouse, S.L., Hairston, N.G. & Cobey, A.E. (1983) Predation and competition in *Ambystoma* larvae: field and laboratory experiments. *J. Herpetol.* **17**: 210–220.

Steward, J.W. (1969) *The Tailed Amphibians of Europe*, Newton Abbot: David & Charles.

Swan, M.J.S. & Oldham, R.S. (1993) *Herptile Sites. Volume 1: National Amphibian Survey Final Report.* English Nature Research Report No. 38. Peterborough: English Nature.

Swierad, J. (1980) Vertical ranges of four species of newts in West Beskid Mountains (Carpathian Mountains, South Poland). *Acta Biol. Cracov. Ser. Zool.* **22**: 35–48.

Szabó, I. (1959) Contributions à l'oecologie de la Salamandre tachetée (*Salamandra salamandra* L.). *Vertebr. Hung.* **1**: 35–48.

Szabó, I. (1962) Nahrungwahl und Nahrung des Gesfleckten Feuersalamanders (*Salamandra salamandra* L.). *Acta Zool. Acad. Sci. Hung.* **8**: 459–457.

Szymura, J.M. (1974) A competitive situation in the larvae of four sympatric species of newts (*Triturus cristatus, T. alpestris, T. montandoni,* and *T. vulgaris*). *Acta Biol. Cracov., Ser. Zool.* **27**: 235–262.

Tarkhnishvili, D.N. & Gokhelashvili, R.K. (1994) Preliminary data of the age structure of a *Mertensiella caucasica* population. *Mertensiella* **4**: 327–334.

Tarkhnishvili, D.N. & Serbinova, I.A. (1993) The ecology of the Caucasian salamander (*Mertensiella caucasica* Waga) in a local population. *Asiat. Herpetol. Res.* **5**: 147–165.

Thiesmeier, B. (1987) Observations on stream drift in larvae of *Salamandra salamandra terrestris* Lacépède, 1788. In van Gelder, J.J., Strijbosch, H. & Bergers, P.J.M. (eds), *Proceedings 4th O.G.M. of Societas Europaea Herpetologica*, pp. 403–406, Nijmegen: Faculty of Sciences.

Thiesmeier, B. (1992) *Ökologie des Feursalamanders*, Essen: Westarp.

Thiesmeier, B. & Hornberg, C. (1990) Zur Fortpflanzung sowie zum Paarungsverhalten der Gebirgsmolche, Gattung *Euproctus* (Gené), im Terrarium, unter besonderer Berücksichtigung von *Euproctus asper* (Dugès, 1852). *Salamandra* **26**: 63–82.

Thonke, A., Jehle, R. & Hödl, W. (1994) Structure, dynamics and phenology of a population of the Danube warty newt (*Triturus dobrogicus*) on the Danube Island near Vienna – a preliminary report. *Abh. Ber. Naturkd.* **17**: 127–133.

Thorn, R. (1968) *Les salamandres d'Europe, Asie et d'Afrique du Nord*, Paris: Lechevalier.

Titus, T.A. & Larson, A. (1995) A molecular phylogenetic perspective on the evolutionary radiation of the salamander family Salamandridae. *System. Biol.* **44**: 125–151.

Trevisan, P. (1982) A new subspecies of alpine salamander. *Boll. Zool.* **49**: 235–239.

Twitty, V.C. (1966) *Of Scientists and Salamanders*, San Francisco: Freeman.

Uiblein, F. (1992) Prey choice behaviour in light and darkness in a facultative cave dweller, the Pyreneen salamander, *Euproctus asper*. *Alytes* **10**: 131–136.

Uiblein, F., Durand, J.P., Juberthie, C. & Parzefall, J. (1992) Predation in caves: the effects of prey immobility and darkness on the foraging behaviour of two salamanders, *Euproctus platycephalus* and *Proteus anguinus*. *Behav. Proc.* **28**: 33–40.

van Gelder, J.J. (1973) Ecological observations on amphibia in the Netherlands. II *Triturus helveticus* Razoumowski: Migration, hibernation and neoteny. *Neth. J. Zool.* **23**: 86–108.

van Koolwijk, T. H. (1987) Collection, trade and protection of European herpetofauna, with special reference to the effects of national and international legislations on the terrarium trade in The Netherlands. Traffic Rapport No. 4.

Veith, M. (1992) The fire salamander, *Salamandra salamandra* (L.), in central Europe: subspecies distribution and intergradation. *Amphibia-Reptilia* **13**: 297–313.

Veith, M. (1994) Morphological, molecular and life history variation in *Salamandra salamandra* (L.). *Mertensiella* **4**: 355–397.

Veith, M. & Dörr, L. (1985) Zur Variabilität morphognostischer Artmerkmale in mitteleuropäischen Teich- und Fadenmolch-Populationen, *Triturus vulgaris vulgaris* (Linnaeus, 1758) und *Triturus helveticus helveticus* (Razoumowski, 1789). *Salamandra* **21**: 197–218.

Vences, M. (1990) Untersuchungen zur Ökologie, Ethologie und geographischen Variation von *Chioglossa lusitanica* Bocage, 1864. *Salamandra* **26**: 267–297.

Verrell, P.A. (1984) Sexual interference and sexual defense in the smooth newt, *Triturus vulgaris* (Amphibia, Urodela, Salamandridae). *Z. Tierpsychol.* **66**: 242–254.

Verrell, P.A. (1985) The emergence of postmetamorphic smooth newts from a pond in southern England. *Br. J. Herpetol.* **6**: 430-432.

Verrell, P.A. (1986) Male discrimination of larger, more fecund females in the mooth newt, *Triturus vulgaris*. *J. Herpetol.* **20**: 416–422.

Verrell, P.A. & Francillon, H. (1986) Body size, age and reproduction in the smooth newt, *Triturus vulgaris*. *J. Zool.* (A) **210**: 89–100.

Verrell, P.A. & Halliday, T.R. (1985) Reproductive dynamics of a population of smooth newts, *Triturus vulgaris*, in southern England. *Herpetologica* **41**: 386–395.

Verrell, P.A., Halliday, T.R. & Griffiths, M.L. (1986) The annual reproductive cycle of the Smooth newt (*Triturus vulgaris*) in England. *J. Zool.* (A) **210**: 101–119.

Vilter, V. & Vilter, A. (1960) Sur la gestation de la salamandre noire des Alpes, *Salamandra atra* Laur. *C. R. Acad. Sci.* Paris **154**: 290–291.

Vilter, A. & Vilter, V. (1963) Mis en évidence d'un cycle reproducteur biennial chez le triton alpestre de montagne. *C. R. Acad. Sci.* Paris **157**: 464–469.

Voesenek, L.A.C.J., van Rooy, P.T.J.C. & Strijbosch, H. (1987) Some autecological data on the urodeles of Sardinia. *Amphibia-Reptilia* **8**: 307–314.

Vorobyeva, E.I. (ed.) (1994) The Siberian newt (*Salamandrella keyserlingii* Dybowski, 1870). Zoogeography, Systematics, Morphology. Russian Academy of Sciences, Moscow. (in Russian).

Wake, D.B. & Özeti, N. (1969) Evolutionary relationships in the family Salamandridae. *Copeia* **1969**: 124–137.

Walhovd, H. (1975) A neotenous female of smooth newt, *Triturus vulgaris. Br. J. Herpetol.* **5**: 510.

Wallis, G.P. & Arntzen, J.W. (1989) Mitochondrial-DNA variation in the crested newt superspecies: limited cytoplasmic gene flow among species. *Evolution* **43**: 88–104.

Warburg, M.R. (1971) The water economy of Israel amphibians: the urodeles *Triturus vittatus* (Jenyns) and *Salamandra salamandra* (L.). *Comp. Biochem. Physiol.* **40A**: 1055–1063.

Warburg, M.R. (1986a) Observations on a relic population of *Salamandra salamandra* on Mt. Carmel during eleven years. In Rocek, Z. (ed.), *Studies in Herpetology*, pp. 389–394, Prague: Charles University.

Warburg, M.R. (1986b) A relic population of *Salamandra salamandra* on Mt. Carmel: a ten year study. *Herpetologica* **42**: 174–179.

Warburg, M.R. (1992) Breeding patterns in a fringe population of fire salamanders, *Salamandra salamandra. Herpetol. J.* **2**: 54–58.

Warburg, M.R. (1994) Population ecology, breeding activity, longevity, and reproductive strategies of *Salamandra salamandra* during an 18-year long study of an isolated population on Mt. Carmel, Israel. *Mertensiella* **4**: 399–421.

Warburg, M.R. & Degani, G. (1979) Evaporative water loss and uptake in juvenile and adult *Salamandra salamandra* (L.) (Amphibia: Urodela). *Comp. Biochem. Physiol.* **62A**: 1071–1075.

Watt, P.J. & Oldham, R.S. (1995) The effect of ammonium nitrate on the feeding and development of larvae of the smooth newt, *Triturus vulgaris* (L.), and on the behaviour of its food source, *Daphnia. Freshwater Biol.* **33**: 319–324.

Zakrzewski, M. (1987) Effect of definite temperature ranges on development, metamorphosis and procreation of the spotted salamander larvae, *Salamandra salamandra* (L.). *Acta Biol. Cracov. Ser. Zool.* **29**: 77–83.

Zuiderwijk, A. (1990) Sexual strategies in the newts *Triturus cristatus* and *Triturus marmoratus. Bijdr. Dierkd.* **60**: 51–64.

Zuiderwijk, A. & Sparreboom, M. (1986) Territorial behaviour in crested newt *Triturus cristatus* and marbled newt *T. marmoratus* (Amphibia, Urodela). *Bijdr. Dierkd.* **56**: 205–213.

Index